The New American Revolution:
The Dawning of the Technetronic Era

The New American Revolution:
The Dawning of the Technetronic Era

Edited by John P. Rasmussen

Department of History
Stanislaus State College
Turlock, California

John Wiley & Sons, Inc.
New York • London • Sydney • Toronto

Library of Congress Catalog Card Number: 73-180275

ISBN 0-471-70918-2 (cloth) ISBN 0-471-70919-0 (paper)

Printed in the United States of America.

10 9 8 7 6 5 4 3 2 1

Preface

One of the greatest challenges confronting the college teacher of American history today is the problem of concluding his courses in an effective and satisfying manner. In the teaching of general surveys, in courses on twentieth century United States, and in classes on American intellectual and social history, it has become exceedingly difficult to wind up these courses and tie things together with any sort of summation. Is there any way to make sense out of our turbulent contemporary scene? It hardly seems adequate to leave the student with little more than a list of catastrophes and a vision of upheaval, pollution, confusion, and doom. Yet how can one present a meaningful, coherent interpretation of an age characterized by such rapid, often chaotic, social change? Also, in trying to tie things together at the end of a course, in attempting to summarize and synthesize, one faces the problem of "relevance." How is the college history teacher to bring in the "now" issues—ecology, the ghetto, automation, the influence of electronic media, the population explosion, etc.? And perhaps the ultimate question for the teacher of history: how does one make the very subject of American history itself relevant to the Now Generation? This book is designed to furnish at least a partial solution to these problems.

The New American Revolution provides a critical examination of the theory that the United States has become the world's first Post-industrial Society. According to this theory, much of the uproar and turbulence we see in America today results from massive, rapid social change which, in turn, is being produced by the impact of advanced technology. The interaction of this massive change and this accelerating technology, it is argued, has produced a truly revolutionary situation: the Post-industrial, Technetronic, or Super-industrial Revolution. This hypothesis has become increasingly fashionable in recent years; it has cropped up in scholarly works, in news magazines, and in at least one presidential message to Congress (*New York Times*, Sept. 12, 1970, p. 8:2). This concept is significant, not merely as a modish way of viewing our society, but also as a highly effective device for the teaching of American history.

v

After three years of extensive classroom experimentation, I am convinced that the Post-industrial Revolution hypothesis provides a most useful means of concluding American history courses in a stimulating manner. As it is a provocative, highly controversial interpretation, this concept is eminently productive of student discussion. It encourages both teacher and student to examine our society and our past from a new perspective. The Post-industrial thesis can serve quite usefully as a core or center around which to focus lectures and discussions on such "instant relevance" issues as pollution, the megalopolis, cybernation, and the entire question of the control and direction of technology. It can also give a frame of reference for reevaluating the nature of our relationships with the other industrial states and with the Third World nations. Finally, the Post-industrial Society hypothesis is quite pertinent to what appears to be one of the newest academic fads: courses analyzing "The Future."

The theme of this book, then, is the relationship between advanced technology and rapid social change—and the relevance of history to that relationship. The essays have been selected to illustrate the controversial nature of the topic and to indicate the wide range of issues and problems involved. The concepts and interpretations included in this book have been thoroughly tested in college classroom use. A short, selective bibliography has been included as a guide for further reading.

I am under obligation to Professors Loren Baritz and Walter Le Feber for their criticisms and suggestions. Any shortcomings of the book are, of course, wholly my responsibility. I am also grateful to many of my students who loyally contributed ideas or suggested articles. I would like to thank the members of the John Wiley & Sons organization for their help and courtesy. I am particularly indebted to Carl Beers for his patience and his expert editorial assistance. Gratitude is also due Miss Barbara Newman, who read the manuscript with meticulous care, and Mr. William R. Hoyt, who gave encouragement when it was needed. I must also thank my loyal typist, Mrs. Fern Anderson, and my wife, who sought, gallantly, persistently, and occasionally successfully to keep our offspring off my neck.

Turlock, California   John P. Rasmussen
December 1971

Contents

The New American Revolution:
The Dawning of the Technetronic Era

During the past ten years, from the end of the cold war and the launching of the first Sputnik, American power has made an unprecedented leap forward. It has undergone a violent and productive internal revolution. Technological innovation has now become the basic objective of economic policy. In America today the government official, the industrial manager, the economics professor, the engineer, and the scientist have joined forces to develop coordinated techniques for integrating factors of production. These techniques have stimulated what amounts to a permanent industrial revolution.

J.-J. Servan-Schreiber●

Introduction

Beyond even the shock of political assassination and the pervasive malaise caused by Viet Nam, Americans found the 1960's and early 1970's a singularly frustrating period. Prosperity and progress, the twin wellsprings of American mythology, had somehow turned bitter. Many intellectuals, students, and minority groups—profoundly alienated from the American Dream—complained loudly about the evils of the military-industrial-scientific complex, the massive pollution of sea and sky, the inequalities of ghetto life, and the various festering crises that affected all levels of the educational system. Outraged radicals insisted that the problems confronting society were unprecedented.

Even to the staid members of Middle America, the national scene was distinctly unsettled and unsettling. Rapid upheaval seemed to render familiar landmarks and guideposts unreliable. The congestion of the freeways was now matched by the congestion of the jet-choked airways. Dominating the American scene was megalopolis, the sprawling super-city whose suburbs defaced the countryside even as its "inner city" rotted into a slum. The international scene was even less reassuring, distinguished as it was by "developing" nations which did not develop—except in terms of soaring populations. In such a world, national security seemed to require complex weapons systems which were astronomical in cost and apocalyptic in destructiveness. Even the exhilaration generated by the United States moon landings—so refreshing after the Soviet space triumphs—was tempered by the immense pricetag of such ventures.

It would seem that there was at least one point upon which both radical and conservative, revolutionary and "square" could agree. The nation was and would continue to be in the grip of *change*—constant, all-encompassing change. The really challenging question was whether there was any pattern, any basic structure, which gave meaning and significance to the unsettling change and the manifold crises which America was undergoing. Was it possible that the various problems faced by the United States were merely symptoms of a deeper, more basic kind of change? And if so, did the fundamental, underlying change represent a break with the American past, or was there a recognizable continuity? Did the upheavals of the 1960's constitute a fork in the road or just a gradual bend in the path of America's social development? What did it all mean? What was it all about?

To the student of American history, the problem of relating the disturbing present to the enigmatic past was not an easy one. Was it possible to fit the current uproar into the more conventional ways of interpreting the nation's past? Or was there now an element of discontinuity which rendered the usual interpretaions of American history—some would say even history itself—quite obsolete and irrelevant? Heretofore, historians have used various theories to explain the way in which our society has developed. The Puritan inheritance and the Protestant ethic, the frontier and virgin land, agrarianism versus capitalism, the absence of feudal tradition and the presence of economic abundance, immigration and ethnic pluralism, and finally geographical and social mobility have all been invoked to demonstrate what America is all about. However, this question persists. Do any of these interpretations aid us in understanding our country *today* in the midst of continuous, revolutionary change?

One explanation seemed refreshingly up to date. Drawing upon studies of the underdeveloped world and of the problem of economic growth, some writers asserted that the most important difference separating the various peoples of the world was the immense gulf between the industrial and the nonindustrial states.

2

Thus, what is most significant about America is that which it shares)
with Europe, Russia, and Japan. It is a developed society, which)
belongs to that elite group of nations which has achieved industrialization. Even this answer, however, is rather less than satisfying. It
hardly seems adequate to state that the multiple crises and
complexities facing America are simply carbon copies of issues
facing Europe—particularly if our problems are unprecedented, as
many radicals assert. Somehow the feeling persisted that the
difficulties confronting the United States were more involved and
more advanced than those shared by the other highly-industrialized
countries.

In the last few years, a diverse group of thinkers has come to the
conclusion that America's dilemmas cannot be understood simply by
viewing the United States as one of several similar industrial
societies. Instead, these authors reason that Americans are going
through an experience that is unique in history. According to this
argument, the most significant thing about our nation is that we are
the first to go beyond the Industrial Revolution to a new stage of
socioeconomic development. America has been undergoing a "post-
industrial revolution" which, in turn, has produced the first truly
technological society. As Professor Comer Vann Woodward has
summed up this line of argument, ". . .we are constantly told that
America and Europe are no longer living in the same historical era,
that the United States has already broken through into something
called the 'post-industrial' or the 'technetronic' age and is in daily
confrontation with the unprecedented. In the American experience,
other peoples of the world can read, for better or for worse, what is
in store for them in the future—for most of them a very remote
future—in which Americans are already living."

According to the "post-industrial revolution" theory, only the
United States has evolved an "intellectual technology"—a vast,
institutionalized research and development effort which stimulates
and guides the entire economy and society. The rise of this system
of organized brainpower has in turn produced a great leap forward.
Science and technology, for example, are now growing at an
exponential rate. America has become the first country in which
government, business and the universities have successfully allied to
produce innovation on an orderly, predictable, and totally unprecedented basis. As John K. Galbraith has pointed out, where
ownership of land was the key to power in premodern times, and
where control of capital was the crucial factor during the first
Industrial Revolution, now it is the command of organized intelli-)
gence that is the *sine qua non* of wealth and prestige.)

Thus, America's massive research program has produced a veritable
explosion of knowledge. Integrated data-processing systems,
advanced metallurgical and electronics technologies, self-contained

3

environmental systems for deep sea and deep space exploration, and sophisticated global communications networks are creating new modes of human existence. Such breakthroughs as automation and systems analysis techniques are producing an intellectual revolution. Doubtless the future will see ultra-low cost power generated by rockburning and waterburning energy plants. To a confirmed optimist, the problems which vex America today are merely the growing pains of a dynamic, progressive society which is living on the cutting edge of destiny. Today's ills can and will be solved by tomorrow's technology. The American, through his mastery of research and development, has become the embodiment of Faustian man.

Yet there is a more sombre aspect to this idea of America as the first technetronic society—a view in which the American is less Faust than he is the sorcerer's apprentice who has conjured up a demon he cannot control. Michael Harrington, among others, has argued persuasively that Americans have allowed their society to drift, willy-nilly, before the forces of technology rather than seeking to control and channel those forces toward socially acceptable goals. The results, he states, have frequently been wasteful and destructive. Harrington also raises another disturbing question. What will Americans, to whom work has traditionally been part of the meaning of life, do with themselves and with their enforced leisure when automation has eliminated productive labor? Has man indeed mortgaged his soul to the machine and sold his birthright for a mess of technological pottage? A growing chorus of angry voices insists that the post-industrial system is most efficient when engaged in ravaging the enviornment and polluting the planet, in producing pesticides, atomic wastes, depleted resources, space junk, and carbon monoxide. Mountains of sewage, stinking to high heaven, will be the tombstone which technology erects in memory of a doomed humanity.

If the technocratic, post-industrial society gives cause for alarm in terms of domestic crisis, the outlook in terms of the international scene is no more reassuring. The thought of a world increasingly split into haves and have-nots was dismal enough. Now one can posit a threefold division of the globe. First, there is the Southern hemisphere with its overpopulation, starvation, backwardness, and mounting violence. Second is above the equator containing the second-class states of Europe, industrialized but small in size and incapable of mutual cooperation—hence dependent upon the United States for protection and for research and advanced technology. Third, "Super Nation," America, with its monopoly of advanced technology—wealthy and powerful beyond the dreams of Croesus— but also the most envied and hated society in the world.

Another international problem raised by the accelerating pace of technology concerns the future relationship of the industrial

4

superpowers. How do the societies and economies of the United States, Russia, and Japan interact? Obviously much of the impetus for the rapid growth of science and technology has been generated by the Cold War competition between the Soviet Union and America. Successive Russian triumphs in thermonuclear weaponry, missile and antimissile systems, and space exploration were greeted in America with nervous soul-searching and redoubled effort in those fields. This pattern of action and reaction reached a climax with the successful launching of the first _Sputnik_ in late 1957. During a veritable frenzy of selfcriticism, Americans demanded that education be revamped, that mathematics, engineering, and science be emphasized, and that above all the nation must be satisfied with nothing less than first place in the international science sweepstakes. The Federal government bowed before this outcry and poured forth vast sums to underwrite an amazing variety of research and development projects. This, according to Servan-Schreiber, helped to create the world's first post-industrial society.

As America and Russia vied with one another in developing huge research establishments and in undertaking elaborate scientific projects, various observers began to question where this would all lead. One optimistic theory has it that a technological imperative will force the two nations to evolve into a similar pattern. Thus the United States and Russia will "converge," rather than collide, and as their social, economic, and technological systems become more and more similar, their political systems will perforce also converge and the tensions of the Cold War will dissipate. Other analysts, however, see no signs that science and technology are creating any sort of convergence between the two systems and even suggest that continued rivalry for superiority in these fields will generate even greater hostility.

The technological revolution produced similar questions concerning the future role of Japan. Following the devastation of World War II, the Japanese people rebuilt their economy with a fanatical drive. Now that the Japanese have created an industrial system of impressive efficiency and achieved the stigmata of a truly developed society (extensive pollution, ugly cities, traffic jams, and rioting students) it remains to be seen what they will do with their creation. Will an industrial giant remain content with a pygmy's role in world affairs? Having the capacity for both a great military establishment, complete with nuclear weapons, and a forceful foreign policy, will Japan not desire these? Or will she choose to continue her passive role in world affairs while devoting all her energies to internal economic development?

At this point, the student of American history must decide for himself whether the concept of the post-industrial revolution and the technocratic society is a valid and useful tool for the interpretation of our culture and national character. Has there, indeed, been a "quantum jump" which has opened a new stage in the Industrial Revolution and has this placed America on a different developmental plane from that occupied by Japan, Russia, and Europe? Is this post-industrial technology the root issue which has given rise to the various crises which confront us? Are the problems of pollution, alienation, urban sprawl, impersonalization, the blight of the cities, etc., merely the symptoms of a more profound development—the arrival of a new phase of industrialization? Are our various problems basically no more than the growing pains—hopefully capable of being cured or outgrown—incident upon the arrival of a new level of civilization? Or has there, perhaps, been far too much overemphasis upon America's supposedly revolutionary scientific progress? Is it not possible that our much vaunted technological superiority is as deceptive as the equally vaunted permanent prosperity of the 1920's? On the other hand, are the prophets of ecological doom and technocratic disaster perhaps guilty of massive overstatement and blatant fear mongering?

However the student may choose to answer these questions, the facts of massive change and of equally massive crises remain and must be explained and somehow placed in the context of the American past. Or, is perhaps America's past irrelevant to America's present problems? Yet, however rapid the pace of change may be, can the present ever really be understood unless it is placed against the perspective of history? Here, perhaps, it is well to ponder the question posed at the beginning of the twentieth century by the brilliant Adams brothers, Henry and Brooks. Can America's essentially eighteenth century political machinery display enough flexibility to cope successfully with the forces of modern science and technology?

"I apprehend for the next hundred years an ultimate, colossal, cosmic collapse; but not on any of our old lines. My belief is that science is to wreck us, and that we are like monkeys monkeying with a loaded shell; we don't in the least know or care where our practically infinite energies come from or will bring us to."

Henry Adams to Brooks Adams, 1902●

I. The Idea of Post-Industrial Society

Increasingly, observers of the American scene are concluding that the assorted ills which beset our society are neither random nor unconnected. From student rebellion to pollution and from urban blight to population explosion, these problems are all related to an unprecedented acceleration of technology. Professor Arthur Schlesinger Jr., of the City University of New York, endeavors to place the various aspects of this "crisis of confidence" in historical perspective. Mr. Schlesinger stands as an eminent authority on United States history who is also deeply involved in contemporary affairs. A leading expert on the reform movements of the Jackson and New Deal eras, he served as "action intellectual" for President John F. Kennedy and is passionately committed to liberal causes and Democratic politics. If the American past can be used to explain the American present and if history can illuminate the problem of the "Post-Industrial Revolution", then Professor Schlesinger would seem well qualified for the task.

●Source: W. C. Ford, ed., *Letters of Henry Adams,* Houghton Mifflin Co. (Boston, 1930).

Arthur M. Schlesinger, Jr. on the Velocity of History*

America is unquestionably experiencing an extreme crisis of confidence. And this crisis is unquestionably not illusory, even if it coexists with affluence, social gains and scientific miracles. It is a crisis with many sources; but none is more important, I think than the incessant and irreversible increase in the rate of social change.

Henry Adams was the first American historian to note the transcendant significance of the ever accelerating velocity of history. "The world did not [just] double or treble its movement between 1800 and 1900," he wrote in 1909, "but, measured by any standard known to science—by horsepower, calories, volts, mass in any shape—the tension and vibration and so-called progression of society were fully a thousand times greater in 1900 than in 1800." Hurried on by the cumulative momentum of science and technology, the tension and vibration of society in 1970 are incalculably greater than they were when Adams wrote. Nor, pending some radical change in human values or capabilites, can we expect the rate of change to slow down at all in the foreseeable future.

This increase in the velocity of history dominates all aspects of contemporary life. First of all, it is responsible for the unprecedented instability of the world in which we live. Science and technology make, dissolve, rebuild and enlarge our environment every week; and the world alters more in a decade than it used to alter in centuries. This has meant the disappearance of familiar landmarks and guideposts that stabilized life for earlier generations. It has meant that children, knowing how different their own lives will be, can no longer look to parents as models and authorities. Change is always scary; uncharted, uncontrolled change can be deeply demoralizing. It is no wonder that we moderns feel forever disoriented and off balance; unsure of our ideas and institutions; unsure of our relations to others, to society and to history; unsure of our own purpose and identity.

The onward roar of science and technology has other consequences. A second springs from the paradox that the very machinery civilization has evolved to create abundance for the mass also creates anxiety for the individual. The high-technology society is above all the society of the great organization. In advanced nations, the great organizations—of government, of industry, of

*Source: Arthur M. Schlesinger Jr., "The Velocity of History," copyright *Newsweek,* Inc., July 6, 1970, pp. 32-34.

education, of research, of communications, of transport, of labor, of marketing—become the units of social energy. These great organizations, as Professor Galbraith points out, generate a life, world and truth of their own; in all areas, the sovereignty of the consumer begins to yield to the sovereignty of the producer; and, in the shadow of these towering structures, the contemporary individual feels puny and helpless. Indeed, no social emotion is more widespread today than the conviction of personal powerlessness, the sense of being beset, beleaguered and persecuted. It extends not only to Black Panthers and members of the Students for a Democratic Society but also to businessmen, publishers, generals and (as we have recently come to observe) Vice Presidents.

A third consequence springs from the technological process itself. For the inner logic of science and technology is rushing us on from the mechanical into the electronic age—into the fantastic new epoch of electronic informational systems and electronic mechanisms of control, the age foreshadowed by television and the computer. One need not be a devout McLuhanite to recognize the force of Marshall McLuhan's argument that history is profoundly affected by changes in the means of communication. There can be little doubt that the electronic age will have penetrating effects not just on the structure and processes of society but on the very reflexes of individual perception. Already television, by its collectiveness and simultaneity, has fostered an intense desire for political self-expression and visibility, as it has spread the habit of instant reaction and stimulated the hope of instant results.

The accelerating velocity of history has, of course, a multitude of other effects. Improved methods of medical care and nutrition have produced the population crisis; and the growth and redistribution of population have produced the urban crisis. The feverish increase in the gross national product first consumes precious natural resources and then discharges filth and poison into water and air; hence the ecological crisis. Nor can one omit the extraordinary moral revolution which makes our comtemporary society reject as intolerable conditions of poverty, discrimination and oppression that mankind had endured for centuries.

All these factors, as they have risen in intensity and desperation, are placing increasing strain on inherited ideas, institutions and values. The old ways by which we reared, educated, employed and governed our people seem to make less and less sense. Half a century ago Henry Adams wrote, "Every historian—sometimes unconsciously but always inevitably—must have put to himself the question: How long . . . could an outworn social system last?" This is a question which historians—and citizens—must consider very intently today.

If this analysis is correct, then the crisis we face is a good deal deeper than simply the anguish over the ghastly folly of Vietnam. For that matter, it is a good deal deeper than is imagined by those who trace all iniquities to the existence of private profit and corporate capitalism. For the acceleration of social change creates its problems without regard to systems of ownership or ideology. The great organization, for example, dominates Communist as much as it does democratic states. Indeed, more so; for the more centralized the ownership and the more absolutist the ideology, the greater the tyranny of organization. The individual stands his better chance in societies where power is reasonably distributed and diversified and where ideas are in free and open competition.

If the crisis seems today more acute in the United States than anywhere else, it is not because of the character of our economic system; it is because the revolutions wrought by science and technology have gone farther here than anywhere else. As the nation at the extreme frontier of technological development, America has been the first to experience the unremitting shock and disruptive intensity of accelerated change. The crises we are living through are the crises of modernity. Every nation, as it begins to reach a comparable state of technical development, will have to undergo comparable crises.

Is the contemporary crisis more profound than other crises in our history? This seems to me an unanswerable question, because a crisis surmounted always seems less terrible than a crisis in being. One cannot say that we are in more danger today than during the Civil War, for example, or during the Great Depression, or during the second world war. One can only say that we met the earlier challenges—and have not yet met this one. And one can add that in earlier crises in our history the quality of our national leadership seemed much more adequate to the magnitude of the challenge.

What Lincoln brought to the Civil War, what Franklin Roosevelt brought to the Great Depression and the second world war, were, above all, an intense imaginative understanding of the nature of the crisis, perceived in the full sweep of history, with a bold instinct for innovation and a determination to mobilize and apply social intelligence and a compassionate sense of human tragedy. It was this embracing vision that gave particular policies their meaning and strength and inspired the American people to rise to their obligations and opportunities. Today, alas, our national leadership hardly seems aware of the fact we are in a crisis; in fact, it hardly appears to know what is going on in America and the world. It is feeble and frightened, intellectually mediocre, devoid of elevation and understanding, fearful of experiment, without a sense of the past or a sense of the future. This is one reason why our crisis of confidence is so acute. It is as if James Buchanan were fighting the Civil War or Herbert Hoover the Great Depression.

The basic task is to control and humanize the forces of change in order to prevent them from tearing our society apart. Our nation is in a state of incipient fragmentation; and the urgent need (along with ending the war) is a national reconstruction that will bring the estranged and excluded groups into full membership in our national community. This means social justice as well as racial justice; it means a far broader measure of participation in our great organizations and institutions; it means the determination to enable all Americans to achieve a sense of function, purpose and potency in our national life. And it means leadership in every area which greatly and generously conceives its responsibility and its hope.

I do not accept the thesis of the inexorable decline of America. No one can doubt that our nation is in trouble. But the present turmoil may be less the proof of decay than the price of progress. The cutting edge of science and technology has sliced through ancient verities and accustomed institutions; it has raised up new questions, new elites, new insistences, new confrontations. As the process has gathered momentum, the immediate result has been—was almost bound to be—social and moral confusion, frustration, fear, violence. Yet the turmoil, the confusion, even the violence may well be the birth pangs of a new epoch in the history of man. If we can develop the intelligence, the will and the leadership to absorb, digest and control the consequences of accelerated technological change, we can avoid the fate of internal demoralization and disintegration—and perhaps offer an example that will help other nations soon to struggle with similar problems and will, in time, restore American influence in the world.

But, as Herbert Croly wrote 60 years ago, we can no longer conceive the promise of American life "as a consummation which will take care of itself . . . as destined to automatic fulfillment." We face a daunting task—still not a bad one for all that. Emerson said, "if there is any period one would desire to be born in—is it not the era of revolution when the old and the new stand side by side and admit of being compared; when all the energies of man are searched by fear and hope; when the historic glories of the old can be compensated by the rich possibilities of the new era? This time like all times is a very good one if one but knows what to do with it."

Zbigniew Brzezinski, political scientist, "Kremlinologist," and former member of the State Department's prestigious ` Policy Planning Council, examines the significance of the post-industrial revolution from a viewpoint somewhat removed from that of Professor Schlesinger—namely, that of an expert on international relations. Although the general tone of this essay is optimistic, it differs from the preceding article in that it raises a number of genuinely frightening possibilities about the future. Among the disturbing questions which Professor Brzezinski brings into focus are the following:

1. Does the coming of the "technetronic age" really mark a new stage in human development, distinct rrom the Industrial Revolution as well as from earlier stages of history—a quantum jump rather than a mere acceleration of technological change?

2. Can our traditional concepts of political democracy and individual liberty survive the coming reign of science and technology, or will the coming of the year 2000 witness the triumph of some kind of "technocratic dictatorship"?

3. Can higher education maintain its independence and objectivity, or will the colleges and universities regress into industrialized, programmed "think-tanks"?

4. If the world is in fact dividing into three spheres, underdeveloped, developed, and superdeveloped, is there any hope for peace and the brotherhood of man, or is the globe doomed to an inevitable series of wars between the various have and have-not powers?

Zbigniew Brzezinski on the Technetronic Era[*]

THE ONSET OF THE TECHNETRONIC AGE

The impact of science and technology on man and his society, especially in the more advanced countries of the world, is becoming the major source of contemporary change. Recent years have seen a proliferation of exciting and challenging leterature on the future. In the United States, in Western Europe, and, to a lesser degree, in Japan and in the Soviet Union, a number of systematic, scholarly efforts have been made to project, predict, and grasp what the future holds for us.

The transformation that is now taking place, especially in America, is already creating a society increasingly unlike its industrial predecessor. The post-industrial society is becoming a "technetronic" society: a society that is shaped culturally, psychologically, socially, and economically by the impact of technology and electronics—particularly in the area of computers and communications. The industrial process is no longer the principal determinant of social change, altering the mores, the social structure, and the values of society. In the industrial society technical knowledge was applied primarily to one specific end: the acceleration and improvement of production techniques. Social consequences were a later by-product of this paramount concern. In the technetronic society scientific and technical knowledge, in addition to enhancing production capabilities, quickly spills over to affect almost all aspects of life directly. Accordingly, both the growing capacity for the instant calculation of the most complex interactions and the increasing availability of biochemical means of human control augment the potential scope of consciously chosen direction, and thereby also the pressures to direct, to choose, and to change.

Reliance on these new techniques of calculation and communication enhances the social importance of human intelligence and the immediate relevance of learning. The need to integrate social change is heightened by the increased ability to decipher the patterns of change; this in turn increases the significance of basic assumptions concerning the nature of man and the desirability of one or another form of social organization. Science thereby

13

intensifies rather than diminishes the relevance of values, but it demands that they be cast in terms that go beyond the more crude ideologies of the industrial age.

New Social Patterns

For Norbert Wiener, "the locus of an earlier industrial revolution before the main industrial revolution" is to be found in the fifteenth-century research pertaining to navigation (the nautical compass), as well as in the development of gunpowder and printing. Today the functional equivalent of navigation is the thrust into space, which requires a rapid computing capacity beyond the means of the human brain; the equivalent of gunpowder is modern nuclear physics, and that of printing is television and long-range instant communications. The consequence of this new technetronic revolution is the progressive emergence of a society that increasingly differs from the industrial one in a variety of economic, political, and social aspects. The following examples may be briefly cited to summarize some of the contrasts:

1. In an industrial society the mode of production shifts from agriculture to industry, with the use of human and animal muscle supplanted by machine operation. In the technetronic society industrial employment yields to sevices, with automation and cybernetics replacing the operation of machines by individuals.

2. Problems of employment and unemployment—to say nothing of the prior urbanization of the post-rural labor force—dominate the relationship between employers, labor, and the market in the industrial society, and the assurance of minimum welfare to the new industrial masses is a source of major concern. In the emerging new society questions relating to the obsolescence of skills, security, vacations, leisure, and profit sharing dominate the relationship, and the psychic well-being of millions of relatively secure but potentially aimless lower-middle-class blue-collar workers becomes a growing problem.

3. Breaking down traditional barriers to education, and thus creating the basic point of departure for social advancement, is a major goal of social reformers in the industrial society. Education, available for limited and specific periods of time, is initially concerned with overcoming illiteracy and subsequently with technical training, based largely on written, sequential reasoning. In the technetronic society not only is education universal but advanced training is available to almost all who have the basic talents, and there is far greater emphasis on quality selection. The essential problem is to

discover the most effective techniques for the rational exploitation of social talent. The latest communication and calculating techniques are employed in this task. The educational process becomes a lengthier one and is increasingly reliant on audio-visual aids. In addition, the flow of new knowledge necessitates more and more frequent refresher studies.

4. In the industrial society social leadership shifts from the traditional rural-aristocratic to an urban-plutocratic elite. Newly acquired wealth is its foundation, and intense competition the outlet—as well as the stimulus—for its energy. In the technetronic society plutocratic pre-eminence is challenged by the political leadership, which is itself increasingly permeated by individuals possessing special skills and intellectual talents. Knowledge becomes a tool of power and the effective mobilization of talent an inportant way to acquire power.

5. The university in an industrial society—in contrast to the situation in medieval times—is an aloof ivory tower, the repository of irrelevant, even if respected, wisdom, and for a brief time the fountain head for budding members of the established social elite. In the technetronic society the university becomes an intensely involved "think tank," the source of much sustained political planning and social innovation.

6. The turmoil inherent in the shift from a rigidly traditional rural society to an urban one engenders an inclination to seek total answers to social dilemmas, thus causing ideologies to thrive in the industrializing society. (The American exception to this rule was due to the absence of a feudal tradition, a point well developed by Louis Hartz.) In the industrial age literacy makes for static interrelated conceptual thinking, congenial to ideological systems. In the technetronic society audio-visual communications prompt more changeable, disparate views of reality, not compressible into formal systems, even as the requirements of science and the new computative techniques place a premium on mathematical logic and systematic reasoning. The resulting tension is felt most acutely by scientists, with the consequence that some seek to confine reason to science while expressing their emotions through politics. Moreover, the increasing ability to reduce soical conflicts to quantifiable and measurable dimensions reinforces the trend toward a more pragmatic approach to social problems, while it simultaneously stimulates new concerns with preserving "humane" values.

7. In the industrial society, as the hitherto passive masses become active there are intense political conflicts over such matters as disenfranchisement and the right to vote. The issue of political participation is a crucial one. In the technetronic age the question is increasingly one of ensuring real participation in decisions that seem too complex and too far removed from

the average citizen. Political alienation becomes a problem. Similarly, the issue of political equality of the sexes gives way to a struggle for the sexual equality of women. In the industrial society woman—the operator of machines—ceases to be physically inferior to the male, a consideration of some importance in rural life, and begins to demand her political rights. In the emerging technetronic society automation threatens both males and females, intellectual talent is computable, the "pill" encourages sexual equality, and women begin to claim complete equality.

8. The newly enfranchised masses are organized in the industrial society by trade unions and political parties and unified by relatively simple and somewhat ideological programs. Moreover, political attitudes are influenced by appeals to nationalist sentiments, communicated through the massive increase of newspapers employing, naturally, the readers' national language. In the technetronic society the trend seems to be toward aggregating the individual support of millions of unorganized citizens, who are easily within the reach of magnetic and attractive personalities, and effectively exploiting the latest communication techniques to maipulate emotions and control reason. Reliance on television—and hence the tendency to replace language with imagery, which is international rather than national, and to include war coverage or scenes of hunger in places as distant as, for example, India—creates a somewhat more cosmopolitan, though highly impressionistic, involvement in global affairs.

9. Economic power in the early phase of industrialization tends to be personalized, by either great entrepreneurs like Henry Ford or bureaucratic industrial officials like Kaganovich, or Minc (in Stalinist Poland). The tendency toward depersonalization of economic power is stimulated in the next stage by the appearance of highly complex interdependence between governmental institutions (including the military), scientific establishments, and industrial organizations. As economic power becomes inseparably linked with political power, it becomes more invisible and the sense of individual futility increases.

10. In an industrial society the acquisition of goods and the accumulation of personal wealth become forms of social attainment for an unprecedentedly large number of people. In the technetronic society the adaptation of science to humane ends and a growing concern with the quality of life become both possible and increasingly a moral imperative for a large number of citizens, especially the young.

Eventually, these changes and many others, including some that more directly affect the personality and quality of the human being himself, will make the technetronic society as different from the industrial as the industrial was from the agrarian. And just as the shift from an agrarian economy and feudal politics toward an industrial society and political systems based on the individual's emotional identification with the nation-state gave rise to contemporary international politics, so the appearance of the technetronic society reflects the onset of a new relationship between man and his expanded global reality.

Social Explosion/Implosion

This new relationship is a tense one: man has still to define it conceptually and thereby render it comprehensible to himself. Our expanded global reality is simultaneously fragmenting and thrusting itself in upon us. The result of the coincident explosion and implosion is not only insecurity and tension but also an entirely novel perception of what many still call international affairs.

Life seems to lack cohesion as environment rapidly alters and human beings become increasingly manipulable and malleable. Everything seems more transitory and temporary: external reality more fluid than solid, the human being more synthetic than authentic. Even our senses perceive an entirely novel "reality"—one of our own making but nevertheless, in terms of our sensations, quite "real." More important, there is already widespread concern about the possibility of biological and chemical tampering with what has until now been considered the immutable essence of man. Human conduct, some argue, can be predetermined and subjected to deliberate control. Man is increasingly acquiring the capacity to determine the sex of his children, to affect through drugs the extent of their intelligence, and to modify and control their personalities. Speaking of a future at most only decades away, an experimenter in intelligence control asserted, "I foresee the time when we shall have the means and therefore, inevitably, the temptation to manipulate the behavior and intellectual functioning of all the people through environmental and biochemical manipulation of the brain."

Thus, it is an open question whether technology and science will in fact increase the options open to the individual. Under the headline "Study Terms Technology a Boon to Individualism," *The New York Times* reported the preliminary conclusions of a Harvard project on the social significance of science. Its participants were quoted as concluding that "most Americans have a greater range of personal choice, wider experience and more highly developed sense of self-worth than ever before." This may be so, but a judgment of this sort rests essentially on an intuitive—and comparative—insight into the present and past states of mind of Americans. In this connection a word of warning from an acute observer is highly relevant: "It behooves us to examine carefully the degree of validity, as measured by actual behavior, of the statement that a benefit of technology will be to increase the number of options and alternatives the individual can choose from. In principle, it could; in fact, the individual may use any number of psychological devices to aviod the discomfort of information overload, and thereby keep the range of alternatives to which he responds much narrower than that which technology in principle makes available to him." In other words, the real questions are how the individual will exploit the options, to what extent he will be intellectually and psychologically prepared to exploit them, and in what way society as a whole will create a favorable setting for taking advantage of these options. Their availability is not of itself proof of a greater sense of freedom or self-worth.

Instead of accepting himself as a spontaneous given, man in the most advanced societies may become more concerned with conscious self-analysis according to external, explicit criteria: What is my IQ? What are my aptitudes, personality traits, capabilities, attractions, and negative features? The "internal man"—spontaneously accepting his own spontaneity—will more and more be challenged by the "external man"—consciously seeking his self-conscious image; and the transition from one to the other may not be easy. It will also give rise to difficult problems in determining the legitimate scope of social control. The possibility of extensive chemical mind control, the danger of loss of individuality inherent in extensive transplantation, the feasibility of manipulating the genetic structure will call for the social definition of common criteria of use and restraint. As the previously cited writer put it, ". . . while the chemical affects the individual, the person is significant to himself and to society in his *social* context—at work, at home, at play. The consequences are social consequences. In deciding how to deal with such alterers of the ego and of experience (and consequently alterers of the personality after the experience), and in deciding how to deal with the 'changed' human beings. we will have to face new questions such as 'Who am I?' 'When am I who?' 'Who are *they* in relation to me?' "

Moreover, man will increasingly be living in man-made and rapidly man-altered environments. By the end of this century approximately two-thirds of the people in the advanced countries will live in cities. Urban growth has so far been primarily the by-product of accidental economic convenience, of the magnetic attraction of population centers, and of the flight of many from rural poverty and exploitation. It has not been deliberately designed to improve the quality of life. The impact of "accidental" cities is already contributing to the depersonalization of individual life as the kinship structure contracts and enduring relations of friendship become more difficult to maintain. Julian Huxley was perhaps guilty of only slight exaggeration when he warned that "overcrowding in animals leads to distorted neurotic and downright pathological behavior. We can be sure that the same is true in principle of people. City life today is definitely leading to mass mental disease, to growing vandalism and possible eruptions of mass violence."

The problem of identity is likely to be complicated by a generation gap, intensified by the dissolution of traditional ties and values derived from extended family and enduring community relationships. The dialogue between the generations is becoming a dialogue of the deaf. It no longer operates within the conservative-liberal or nationalist-internationalist framework. The breakdown in communication between the generations—so vividly evident during the student revolts of 1968—was rooted in the irrelevance of the old symbols to many younger people. Debate implies the acceptance of a common frame of reference and language; since these were lacking, debate became increasingly impossible.

Though currently the clash is over values—with many of the young rejecting those of their elders, who in turn contend that the young have evaded the responsibility of articulating theirs—in the future the clash between generations will be also over expertise. Within a few years the rebels in the more advanced countries who today have the most visibility will be joined by a new generation making its claim to power in government and business: a generation trained to reason logically; as accustomed to exploiting electronic aids to human reasoning as we have been to using machines to increase our own mobility; expressing itself in a language that functionally relates to these aids; accepting as routine managerial processes current innovations such as planning-programming-budgeting systems (PPBS) and the appearance in high business echelons of "top computer executives." As the older elite defends what it considers not only its own vested interests but more basically its own way of life, the resulting clash could generate even more intense conceptual issues.

Global Absorption

But while our immediate reality is being fragmented, global reality increasingly absorbs the individual, involves him, and even occasionally overwhelms him. Communications are the obvious, already much discussed, immediate cause. The changes wrought by communications and computers make for an extraordinarily interwoven society whose members are in continuous and close audio-visual contact—constantly interacting, instantly sharing the most intense social experiences, and prompted to increased personal involvement in even the most distant problems. The new generation no longer defines the world exclusively on the basis of reading, either of ideologically structured analyses or of extensive descriptions; it also experiences and senses it vicariously through audio-visual communications. This form of communicating reality is growing more rapidly—especially in the advanced countries—than the traditional written medium, and it provides the principal source of news for the masses. . . "By 1985 distance will be no excuse for delayed information from any part of the word to the powerful urban nerve centers that will mark the major concentrations of the people on earth." Global telephone dialing that in the more advanced states will include instant visual contact and global television-satellite system that will enable some states to "invade" private homes in other countries will create unprecedented global intimacy.

The new reality, however, will not be that of a "global village." McLuhan's striking analogy overlooks the personal stability, interpersonal intimacy, implicitly shared values, and traditions that were important ingredients of the primitive village. A more appropriate analogy is that of the "global city"—a nervous, agitated, tense, and fragmented web of interdependent relations. That interdependence, however, is better characterized by interaction than by intimacy. Instant communications are already creating something akin to a global nervous system. Occasional malfunctions of this nervous system—because of blackouts or breakdowns—will be all the more unsettling, precisely because the mutual confidence and reciprocally reinforcing stability that are characteristic of village intimacy will be absent from the process of that "nervous" interaction.

Man's intensified involvement in global affairs is reflected in, and doubtless shaped by, the changing character of what has until now been considered local news. Television has joined newspapers in expanding the immediate horizons of the viewer or reader to the point where "local" increasingly means "national," and global affairs compete for attention on an unprecedented scale. Physical and moral immunity to "foreign" events cannot be very effectively maintained under circumstances in which there are both a growing intellectual awareness of global interdependence and the electronic intrusion of global events into the home.

This condition also makes for a novel perception of foreign affairs. Even in the recent past one learned about international politics through the study of history and geography, as well as by reading newspapers. This contributed to a highly structured, even rigid, approach, in which it was convenient to categorize events or nations in somewhat ideological terms. Today, however, foreign affairs intrude upon a child or adolescent in the advanced countries in the form of disparate, sporadic, isolated—but involving—events: catastrophes and acts of violence both abroad and at home become intermeshed, and though they may elicit either positive or negative reactions, these are no longer in the neatly compartmentalized categories of "we" and "they." Television in particular contributes to a "blurred," much more impressionistic—and also involved—attitude toward world affairs. Anyone who teaches international politics senses a great change in the attitude of the young along these lines.

Such direct global intrusion and interaction, however, does not make for better "understanding" of our contemporary affairs. On the contrary, it can be argued that in some respects "understanding—in the sense of possessing the subjective confidence that one can evaluate events on the basis of some organized principle—is today much more difficult for most people to attain. Instant but vicarious participation in events evokes uncertainty, especially as it becomes more and more apparent that established analytical categories no longer adequately encompass the new circumstances.

The science explosion—the most rapidly expanding aspect of our entire reality, growing more rapidly than population, industry, and cities—intensifies, rather than reduces, these feelings of insecurity. It is simply impossible for the average citizen and even for men of intellect to assimilate and meaningfully organize the flow of knowledge for themselves. In every scientific field complaints are mounting that the torrential outpouring of published reports, scientific papers, and scholarly articles and the proliferation of professional journals make it impossible for individuals to avoid becoming either narrow-gauged specialists or superficial generalists. The sharing of new common perspectives thus becomes more difficult as knowledge expands; in addition, traditional perspectives such as those provided by primitive myths or more recently, by certain historically conditioned ideaologies can no longer be sustained.

The threat of intellectual fragmentation, posed by the gap between the pace in the expansion of knowledge and the rate of its assimilation, raises a perplexing question concerning the prospects for mankind's intellectual unity. It has generally been assumed that the modern world, shaped increasingly by the industrial and urban revolutions, will become more homogeneous in its outlook. This may be so, but it could be the homogeneity of insecurity, of uncertainty, and of intellectual anarchy. The result, therefore, would not necessarily be a more stable environment.

THE AMBIVALENT DISSEMINATOR

The United States is the principal global disseminator of the technetronic revolution. It is American society that is currently having the greatest impact on all other societies, prompting a far-reaching cumulative transformation in their outlook and mores. At various stages in history different societies have served as a catalyst for change by stimulating imitation and adaptation in others. What in the remote past Athens and Rome were to the Mediterranean world, or China to much of Asia, France has more recently been to Europe. French letters, arts, and political ideas exercised a magnetic attraction, and the French Revolution was perhaps the single most powerful stimulant to the rise of populist nationalism during the nineteenth century.

In spite of its domestic tensions—indeed, in some respects because of them (. . .)—the United States is the innovative and creative society of today. It is also a major disruptive influence on the world scene. In fact communism, which many Americans see as the principal cause of unrest, primarily capitalizes on frustrations and aspirations, whose major source is the American impact on the rest of the world. The United States is the focus of global attention, emulation, envy, admiration, and animosity. No other society evokes feelings of such intensity; no other society's internal affairs—including America's racial and urban violence—are scrutinized with such attention; no other society's politics are followed with such avid interest—so much so that to many foreign nationals United States domestic politics have become an essential extension of their own; no other society so massively disseminates its own way of life and its values by means of movies, television, multimillion-copy foreign editions of its national magazines, or simply by its products; no other society is the object of such contradictory assessments.

The American Impact

Initially, the impact of America on the world was largely idealistic: America was associated with freedom. Later the influence became more materialistic: America was seen as the land of opportunity, crassly defined in terms of dollars. Today similar material advantages can be sought elsewhere at lower personal risk, and the assassinations of the Kennedys and of Martin Luther King, as well as racial and social tensions, not to speak of Vietnam, have somewhat tarnished America's identification with freedom. Instead, America's influence is in the first instance scientific and technological, and it is a function of the scientific, technological, and educational lead of the United States.

Scientific and technological development is a dynamic process. It depends in the first instance on the resources committed to it, the personnel available for it, the educational base that supports it, and—last but not least—the freedom of scientific innovation. In all four respects the American position is advantageous; contemporary America spends more on science and devotes greater resources to research than any other society.

In addition, the American people enjoy access to education on a scale greater than that of most other advanced societies (. . . .) At the beginning of the 1960s the United States had more than 66 per cent of its 15-19 age group enrolled in educational institutions; comparable figures for France and West Germany were about 31 per cent and 20 per cent, respectively. The combined populations of France, Germany, Italy, and the United Kingdom are equal to that of the United States—roughly two hundred million. But in the United States 43 per cent of college-age people are actually enrolled, whereas only 7 to 15 per cent are enrolled in the four countries (Italy having the low figure and France the high). The Soviet percentage was approximately half that of the American. In actual numbers there are close to seven million college students in the United States and only about one and a half million in the four European countries. At the more advanced level of the 20-24 age bracket, the American figure was 12 per cent while that for West Germany, the top Western European country, was about 5 per cent. For the 5-19 age bracket, the American and the Western European levels were roughly even (about 80 per cent), and the Soviet Union trailed with 57 per cent.

As a result, the United States possesses a pyramid of educated social talent whose wide base is capable of providing effective support to the leading and creative apex. This is true even though in many respects American education is often intellectually deficient, especially in comparison with the more rigorous standards of Western European and Japanese secondary institutions. Nonetheless, the broad base of relatively trained people enables rapid adaptation, development, and social application of scientific innovation or discovery. While no precise estimates are possible, some experts have suggested that a present-day society would experience difficulties in rapid modernization if less than 10 per cent of its population in the appropriate age bracket had higher education and less than 30 per cent had lower education.

Moreover, both the organizational structure and the intellectual atmosphere in the American scientific world favor experimentation and rapid social adaptation. In a special report on American scientific policies, submitted in early 1968, a group of experts connected with OECD concluded that America's scientific and technical enterprise is deeply rooted in American tradition and history. Competitiveness and the emphasis on quick exploitation have resulted in a quick spinoff of the enormous defense and space research efforts into the economy as a whole, in contrast to the situation in the Soviet Union, where the economic by-products of almost as large-scale a research effort have so far been negligible. It is noteworthy that "the Russians themselves estimate that the productivity of their researchers is only about half the Americans' and that innovations take two or three times as long to be put into effect."

This climate and the concomitant rewards for creative attainments result in a magnetic pull (the "brain drain") from which America clearly benefits. America offers to many trained scientists, even from advanced countries, not only greater material rewards but a unique opportunity for the maximum fulfillment of their talents. In the past Western writers and artists gravitated primarily toward Paris. More recently the Soviet Union and China have exercised some ideological attraction, but in neither case did it involve the movement of significant percentages of scientific elites. Though immigrating scientists initially think of America as a platform for creative work, and not as a national society to which they are transferring political allegiance, in most cases that allegiance is later obtained through assimilation. America's professional attraction for the global scientific elite is without historic precedent in either scale or scope.

Though this attraction is likely to decline for Europeans (particularly because of America's domestic problems and partially because of Europe's own scientific advance), the success of J. J. Servan-Schreiber's book, *The American Challenge,* reflects the basic inclination of concerned Europeans to accept the argument that the United States comes closest to being the only truly modern society in terms of the organization and scale of its economic market, business administration, research and development, and education. (In contrast, the structure of American government is viewed as strikingly antiquated.) European sensitivity in this area is conditioned not only by fear of a widening American technological lead but very much by the increasing presence on the European markets of large American firms that exploit their economic advantages of scale and superior organization to gradually acquire controlling interests in key frontier industries. The presence of these firms, the emergence under their aegis of something akin to a new international corporate elite, the stimulation given by their presence to the adoption of American business practices and training, the deepening awareness that the so-called technology gap is in reality also a management and education gap—all have contributed both to a positive appraisal of American "technostructure" by the European business and scientific elite and to the desire to adapt some of America's experience.

Less tangible but no less pervasive is the American impact on mass culture, youth mores, and life styles. The higher the level of per-capita income in a country, the more applicable seems the term "Americanization." This indicates that the external forms of characteristic contemporary American behavior are not so much culturally determined as they are an expression of a certain level of urban, technical, and economic development. Nonetheless, to the extent that these forms were first applied in America and then "exported" abroad, they became symbolic of the American impact and of the innovation-emulation relationship prevailing between America and the rest of the world.

What makes America unique in our time is that confrontation with the new is part of the daily American experience. For better or for worse, the rest of the world learns what is in store for it by observing what happens in the United States: whether it be the latest scientific discoveries in space and medicine or the electric toothbrush in the bathroom; pop art or LSD; air conditioning or air pollution; old-age problems or juvenile delinquency. The evidence is more elusive in such matters as style, music, values, and social mores, but there too the term "Americanization" obviously implies a specific source.

Similarly, foreign students returning from American universities have prompted an organizational and intellectual revolution in the academic life of their countries. Changes in the academic life of Germany, the United Kingdom, Japan, and more recently France, and to an even greater extent in the less developed countries, can be traced to the influence of American educational institutions. Given developments in modern communications, it is only a matter of time before students at Columbia University and, say, the University of Teheran will be watching the same lecturer simultaneously.

This is all the more likely because American society, more than any other, "communicates" with the entire globe. Roughly [65] per cent of all world communications originate in this country. Moreover, the United States has been most active in the promotion of a global communications system by means of satellites, and it is pioneering the development of a world-wide information grid. It is expected that such a grid will come into being by about 1975. For the first time in history the cumulative knowledge of mankind will be made accessible on a global scale—and it will be almost instantaneously available in response to demand.

New Imperialism?

All of these factors make for a novel relationship between the United States and the world. There are imperial overtones to it, and yet in its essence the relationship is quite different from the traditional imperial structure. To be sure, the fact that in the aftermath of World War II a number of nations were directly dependent on the United States in matters of security, politics, and economics created a system that in many respects, including that of scale, superficially resembled the British, Roman, and Chinese empires of the past. The more than a million American troops stationed on some four hundred major and almost three thousand minor United States military bases scattered all over the globe, the forty-two nations tied to the United States by security pacts, the American military missions training the officers and troops of many other national armies, and the approximately two hundred thousand United States civilian government employees in foreign posts all make for striking analogies to the great classical imperial systems.

Nevertheless, the concept of "imperial" shields rather than reveals a relationship between America and the world that is both more complex and more intimate. The "imperial" aspect of the relationship was, in the first instance, a transitory and rather spontaneous response to the vacuum created by World War II and to the subsequent felt threat from communism. Moreover, it was neither formally structured nor explicitly legitimized. The "empire" was at most an informal system marked by the pretense of equality and noninterference. This made it easier for the "imperial" attributes to recede once conditions changed. By the late 1960s, with a few exceptions, the earlier direct political-military dependence on the United States had declined (often in spite of political efforts by the United States to maintain it). Its place had been filled by the more pervasive but less tangible influence of American economic presence and innovation as they originated directly from the United States or were stimulated abroad by American foreign investment (the latter annually yielding a product considerably in excess of the gross national product of most major countries). In effect, ". . .American influence has a porous and almost invisible quality. It works through the interpenetration of economic institutions, the sympathetic harmony of political leaders and parties, the shared concepts of sophisticated intellectuals, the mating of bureaucratic interests. It is, in other words, something new in the world, and not yet well understood."

It is the novelty of America's relationship with the world—complex, intimate, and porous—that the more orthodox, especially Marxist, analyses of imperialism fail to encompass. To see that relationship merely as the expression of an imperial drive is to ignore the part played in it by the crucial dimension of the technological-scientific revolution. That revolution not only captivates the imagination of mankind (who can fail to be moved by the spectacle of man reaching the moon?) but inescapably compels imitation of the more advanced by the less advanced and stimulates the export of new techniques, methods, and organizational skills from the former to the latter. There is no doubt that this results in an asymmetrical relationship, but the content of that asymmetry must be examined before it is called imperialism. Like every society, America no doubt prefers to be more rather than less advanced; yet it is also striking that no other country has made so great an effort, governmentally and privately, through business and especially through foundations, to export its know-how, to make public its space findings, to promote new agricultural techniques, to improve educational facilities, to control population growth, to improve health care, and so on. All of this has imperial overtones, and yet it is misleading to label it as such.

Indeed, unable to understand fully what is happening in their own society, Americans find it difficult to comprehend the global impact that that society has had in its unique role as disseminator of the technetronic revolution. This impact is contradictory: it both promotes and undermines American interests as defined by American policymakers; it helps to advance the cause of cooperation on a larger scale even as it disrupts existing social or economic fabrics; it both lays the groundwork for well-being and stability and enhances the forces working for instability and revolution. Unlike traditional imperialistic powers, which relied heavily on the principle of *divide et impera* (practiced with striking similarity by the British in India and more recently by the Russians in Eastern Europe), America has striven to promote regionalism both in Europe and in Latin America. Yet in so doing, it is helping to create larger entities that are more capable of resisting its influence and of competing with it economically. Implicitly and often explicitly modeled on the American pattern, modernization makes for potentially greater economic well-being, but in the process it disrupts existing institutions, undermines prevailing mores, and stimulates resentment that focuses directly on the source of change—America. The result is an acute tension between the kind of global stability and order that America subjectively seeks and the instability, impatience, and frustration that America unconsciously promotes.

The United States has emerged as the first global society in history. It is a society increasingly difficult to delineate in terms of its outer cultural and economic boundaries. Moreover, it is unlikely that in the foreseeable future America will cease to exercise the innovative stimulus that is characteristic of its current relationship with the world. By the end of this century (extrapolating from current trends) only some thirteen countries are likely to reach the 1965 level of the per-capita gross national product of the United States. Unless there is major scientific and economic stagnation or a political crisis (. . .), at the end of the century America will still be a significant force for global change, whether or not the dominant subjective mood is pro- or anti-American.

Do the terms "Post-Industrial Society," "Technetronic Era," and "Electronic Revolution" really stand for something new and, presumably, better, or do they merely represent the latest version of the ancient American faith in "progress"—i.e. the destructive and wasteful impact of uncontolled technology upon the hapless American Landscape? Can science and technology be trusted to lead us into the New Jerusalem? What assurance do we have that they will not lead us into the ditch—or the quicksand? Is it not possible that the apotheosis of technology will result in the sterility and stagnation of arts and letters as well as the fatal and irrevocable centralization of political economic, and even psychological power? Using American intellectual history as a framework and equipped with a highly skeptical attitude, Professor James W. Carey, a communications research specialist at University of Illinois, and John J. Quirk, a writer and former student at Illinois, address themselves to these issues.

James W. Carey and John J. Quirk on the Electronic Revolution*

In Thornton Wilder's novel *The Eighth Day,* a typical Illinois town provides the setting for a turn-of-the-century celebration that reflects the anticipations of those Americans who identified change and hope with the coming of the year 1900. Toward the end of the nineteenth century, Americans who had witnessed the destructive effects of industrialization were subject to a naive yearning for a rebirth of native optimism and a resuscitation of the bright promises of science and technology. Wilder's title is taken from the theme of a speech by a community leader who voices the concerns and expectations of those times in words of evolutionary religion. Wilder's speaker envisions the new century as an "eighth day," after Genesis, and men of this century as a new breed, free from the past and heir to the future.

●Source: James W. Carey and John J. Quirk, "The Mythos of the Electronic Revolution," *The American Scholar,* Volume 39, Nos. 2 and 3, Spring and Summer, 1970. Copyright © 1970 by the United Chapters of Phi Beta Kappa. By permission of the publishers.
Note: Deletions have been made in this essay (indicated by asterisks) for the sake of brevity.

In the last third of this century, we are witnessing another prophecy of an "eighth day," punctuated by sophisticated projections of the Year 2000, Mankind 2000, and announcements of the coming of an "Electronic Revolution." In the past, industrial exhibitions and addresses by prominent figures at World's Fairs have been employed to enhance the prestige of technological innovations and to enlist the support of public opinion on behalf of science. Today the Commission on the Year 2000, the World Future Society, and Rand Corporation have become the agencies of prophecy; and the public is invited to participate in such elaborate devices as the "World Future Game" of R. Buckminster Fuller. Nevertheless, the language of contemporary futurology contains an orientation of secular religiosity that surfaces whenever the name of technology is invoked.

This futurist mentality has much in common with the outlook of the Industrial Revolution, which was heralded by Enlightenment philosophers and nineteenth-century moralists as the vehicle of general progress, moral as well as material. Contemporary images of the future also echo the promise of an eighth day and thus predict a radical discontinuity from past history and the present human condition. The dawn of this new era is alternatively termed the "post-industrial society," "post-civilization," "the technetronic society" and "the global village." The new breed of man inhabiting the future is characterized as the "post-modern man," "the protean personality" and "the post-literate-electronic man."

Our topic in this extended essay is an increasingly prevalent and popular brand of the futurist ethos that identifies electricity and electrical power, electronics and cybernetics, with a new birth of community, decentralization, ecological balance and social harmony. This set of notions is most readily associated with Marshall McLuhan, but his status as a celebrity merely obscures his position in a school of thought that has been articulated and reiterated over many decades and that has many spokesmen in our time. The notion of an Electronic Revolution is supported by a diverse consensus that includes designer R. Buckminster Fuller, musicologist John Cage, policy scientist Zbigniew Brzezinski, elements of the New Left, and theologians inspired by Teilhard de Chardin. Outside intellectual circles, the "electronic revolution" and the concepts of McLuhan and others have been repeated and embraced by coteries of advertisers and engineers, corporation and foundation executives, and government personnel.

What brings together this anomalous collection under the banner of the Electronic Revolution is that they are in a real sense the children of the "eighth day," of the millennial impulse resurfacing in response to social crises and technical change. They have cast themselves in the role of secular theologians composing theodicies for electricity and its technological progeny.

Despite the diversity of their backgrounds and positions on other questions, there is within their rhetorical descriptions of the Electronic Revolution a common set of ideas. They all convey an impression that electrical technology is the great benefactor of mankind. Simultaneously, they hail electrical techniques as the motive force of desired social change, the key to the re-creation of a humane community, the means for returning to a cherished naturalistic bliss. Their shared belief is that electricity will overcome historical forces and political obstacles that prevented previous utopias.

Zbigniew Brzezinski pins his view of the future to the belief that: "Ours is no longer the conventional revolutionary era; we are entering a novel metamorphic phase in human history" which is "imposing upon Americans a special obligation to ease the pains of the resulting confrontation" between our society and the rest of the world. In his new version of manifest destiny, Brzezinski suggests that technetronic America will supersede any other social system because all other revolutions have only "scratched the surface. . .alterations in the distribution of power and wealth," while the technetronic revolution will "affect the essence of individual and social existence."

With typically American optimism, Brzezinski enunciates the compatibility of democracy, decentralism and technology. "Yet," Brzezinski continues, "it would be highly misleading to construct a one-sided picture, a new Orwellian piece." "Many of the changes transforming American society augur well for the future." Among those trends Brzezinski identifies "greater devolution of authority" and "massive diffusion of scientific and technical knowledge as a principal focus of American involvement in world affairs" since "technetronics are eliminating the twin insulants of time and space." The resulting situation is one in which a band of social scientists, above party and faction, is enabled to "reduce social conflicts to quantifiable and measurable dimensions, reinforce the trend towards a more pragmatic problem solving approach to social issues."

In McLuhan's scenario it is the Artist rather than the Scientist who is heir to the future. Nonetheless McLuhan dresses electricity in a cloak of mystery as the new invisible hand of providence: "The electronic age, if given its own unheeded lee-way, will drift quite naturally into modes of cosmic humanism. . . ." Far more metaphysical than Brzezinski, McLuhan sees in electricity the capacity to "abolish space and time alike" as it confers the mythic dimension on ordinary industrial and social life today." Finally, McLuhan's penchant for religious metaphors leads to a characterization of elecricity as Divine Force: "The computer, in short, promises by technology a Pentecostal condition of universal understanding and unity."

Whether the rhetoric of the electronic revolution appears in sacred or secular form, it attributes intrinsically benign and progressive properties to electricity and its applications. It also displays a faith that electricity will exorcise social disorder and environmental disruption, eliminate political conflict and personal alienation, and restore ecological balance and a communion of man with nature.

Such a faith, however, contrasts sharply with developments in electricity and electronics in recent decades. The manifest consequences of electricity are clearly in opposition to a decentralized, organic, harmonious order. The use of electronic technology has been biased toward the recentralization of power in computer centers and energy grids, the Pentagon and NASA, General Electric and Commonwealth Edison. Further, the "electronic society" has been characterized by thermal and atmospheric pollution from the generation of electricity and the erosion of regional cultures by television and radio networks the programming of which focuses upon a single national accent in tone and topical coverage at the expense of local idiom interest.

And that leads to a dilemma: either modern electricians possess insight into the future that we are barred from possessing or the revolution announced in their rhetoric is mere wishful thinking, or worse, a new legitimation of the status quo. The latter thought is particularly disturbing, for we may be witnessing the projection into the twenty-first century of certain policies of American politics and industry that have in the past had particularly destructive effects. We also may be mystified concerning a possiblity through the grand eloquence of electrical nomenclature. Electricity is not exactly new, however, and in the history of technology and its social use we may find the terms to appraise the possibilities and potentialities of the electronic revolution.

There is no way to interpret sensibly the claims of electrical utopians except against the background of traditional American attitudes toward technology. For the chastening effect, we should, therefore, remind ourselves of the typical American response to the onset of industrialization and the development of mechanical technology.

America was dreamed by Europeans before it was discovered by Columbus. Atlantis, Utopia, the Passage to India—this land was the redemption of European history before it was the scene of American society. The controlling metaphor that invoked this promised land was Nature, the healing power of an unsullied virgin wilderness. Americans subsequently came to define their "nation's nature" in terms of a pastoral idiom inherited from the European utopians. While mechanical technology was welcomed here, it was to undergo a characterological change when received into the Garden of America. Machinery was to be implanted into and humanized by an idealized rural landscape. The grime, desolation, poverty, injustice and class struggle typical of the European city were not to be reproduced here. America's redemption from European history, her uniqueness, was to be through unblemished nature, which would allow us to have the factory without the factory system, machines without a mechanized society.

A vital and relevant tradition in American studies, inspired by Perry Miller and Henry Nash Smith and continued by Leo Marx and Alan Trachtenburg, has traced the recurrent theme of the "machine in the garden." This was a unique

American idea of a new dimension in social existence through which men might return to an Edenic estate through a harmonious blending of nature and manufactures. Each new invention or device was heralded as a means to move toward the goal of a new environment made possible by the geographical and historical options afforded the young nation. This vision was of a middle landscape, an America suspended between art and nature, between the rural landscape and the industrial city, where technological power and democratic localism could comprise an ideal way of life. As dreamed by intellectuals, preached by ministers, painted by artists, romanticized by politicians, dramatized by novelists, this society was to be located symbolically and literally midway between the overdeveloped nations of Europe and the primitive communities of the western frontier. America's virgin land and abundant resources would produce an indigenous solution to industrialization on this continent, a solution that would rejuvenate all Europeans who ventured into the New World, and would allow us to leap over the disadvantages of the European system of industrialization. America was, in short, exempt from history: from mechanics and industrialization we would derive wealth, power and productivity; from nature—peace, harmony and self-sufficiency.

Influential Enlightenment philosophers anticipated this rhetoric in forecasting the American future. Condorcet, for example, was convinced that America was freed from the dead hand of the past and "would double the progress of the race" and "make it doubly swift." He believed that America was safely insulated from the Old World turmoil, possessed of sufficient space for preservation of rustic virtues, and could translate material progress into moral improvement and social bliss. It was this attitude that converted Jefferson and his agrarian followers to acceptance of the Hamiltonian program of manufactures and infant industry. Jefferson suspended his skepticism about factory economics and came to differentiate between "the great cities in the old countries. . .[where] want of food and clothing. . .[had] begotten a depravity of morals, a dependence and corruption" and America where "manufactures are as much at their ease, as independent and moral as our agricultural inhabitants, and they will continue so long as there are vacant lands for them to resort to."

A special importance was attached periodically to specific technologies that performed key services. Jefferson himself once remarked that newspapers were more necessary than government itself, and he equated the technology of print and the protection of the rights of a free press with literacy and liberty. Patriotic historians even dated the birth of national consciousness with the publication of the first newspaper in Boston in 1704. Finally, the Bill of Rights guaranteed constitutional protection to technology with its clause on freedom of the press.

Later, steam engines occupied a particular place in the pantheon of technologies, for their capacity to link the continent by railroad and waterway and to create new commerical bonds. Eventually there were essays on and oratorical praises of "The Moral Influence of Steam" and "The Indirect Influence of Railroads." A typical passgae of the era, this from an address by Charles Fraser to the Mercantile Library Association of Charleston, South Carolina, invests machinery with metaphysical properties: "An agent was at hand to bring everything into harmonious cooperation. . .triumphing over space and time. . .to subdue prejudice and to unite every part of our land in rapid and friendly communication; and that great motive agent was steam."

Lifting the hyperboles of technological sublimity to a philosophical plane, Emerson paired steam and electromagnetism with transcendentalism: "Machinery and transcendentalism agree well. . .Stage-Coach and Railroad are bursting the old legislation like green withes. . .Our civilization and these ideas are reducing the earth to a brain. See how by telegraph and steam the earth is anthropoligized." In Emerson's aphorism, we have a graphic example of the intellectual's awe of technology and the confusion of technological fact with spiritual symbolism.

The rhetoric of the technological sublime, as Leo Marx has felicitously labeled these tributes to the technology of steam and mechanics, constituted the false consciousness of the decades before the Civil War. Neither the printing press nor the steam engine forestalled that fateful conflict, however, or insured that the victory won by Lincoln and Grant would not be lost during Reconstruction. During the Civil War and in the decades thereafter, the American dream of the mechanical sublime was decisively reversed. It became increasingly evident that America was not exempt from history or isolated from the European experience of a continental democracy. In its aftermath American cities were turned into industrial slums, class and race warfare were everyday features of life, economic stability was continually interrrupted by depression, the countryside was scarred and ravaged by the railroads, coal and iron mining, and the devastation of forests. But reality was unable to reverse rhetoric and in the last third of the nineteenth century, as the dreams of a mechanical utopia gave way to the realities of industrialization, there arose a new school of thought dedicated to the notion that there was a qualitative difference between mechanics and electronics, between machines and electricity, between mechanization and elctrification. In electricity was suddenly seen the power to redeem all the dreams betrayed by the machine.

There were many exemplars of the turn from the mechanical to electrical sublime, but a useful starting figure is the principal American economist of the nineteenth century, Henry Charles Carey. His father, Mathew Carey, an Irish rebel refugee and founding member of the Society for the Promotion of National Industry in Pennsylvania, had published the influential series of writings with which Henry Clay underpinned his "American Plan" for protection of native industries and vast internal improvements in canals and highways. Henry Charles Carey himself rejected Manchester economics and argued for a unique viewpoint in American industrial policy. He suggested that the introduction of the factory and the injection of industry on the native scene would have quite different results in this country than in Europe. Technology on this continent would produce wealth and industrial efficiency but without the wage slavery and environmental disasters of British and European centers.

In 1848, Carey wrote *Past, Present and Future,* a book that formulated a programmatic statement of these ideals in a distinct alternative, "regional associationism." He held that his new system would be realized when regional patterns of "association" between industry and agriculture were founded and merged into cooperative economy. He thought that his plan would permanently secure decentralized, small-scale units in politics as well as economics. In addition, Carey believed that a union of agriculture, industry and universal education in mechanical skills would prevent divisions between country and city and conflicts between social classes.

When Henry Charles Carey was born, Washington was president; in the year he died, Henry Ford began work on motorcars in Michigan. During this life, the idea of regional associationism went unrealized as centralization of industry, money and influence, and the exploitation of immigrant labor, were the overwhelming realities of American life. Carey did not, however, give up the "American Plan" but projected it in the language of electricity. In his last book, *The Unity of Law,* Carey substituted the language of electricity for the language of mechanics, identifying the physical laws of electricity and magnetism, then being discovered, with the laws of society, and projecting electricity as the new bond between nature and society. One dense but important quotation will illustrate this shift:

> Electricity presents a far more striking resemblance to the brain power which is its correspondent in the societary life. So striking indeed is it that when we need to express the idea of rapid action of the societary thought and will, we find ourselves compelled to look to the physical world for the terms to be employed, availing ourselves of those of electricity and magnetism. . .
>
> The actual relation of each and every member of a community as giver and receiver, teacher and learner, producer and consumer is positive and negative by turns and relatively to every difference of function and force in his associates, the whole mass constituting a great electric battery to which each individual contributes his pair of plates. Perfect circulation being established as a consequence of perfect development of all individualities, the economic force flows smoothly through every member of the body politic, general happiness and prosperity, improved mental and moral action following in its train. . .
>
> . . .wealth and power. . .everywhere in the ratio in which each and every pair of plates is placed in proper relation with each other; the

vitalized circuit being thus established throughout the entire mass and made to bear, with the concentrated energy of the whole upon every object of general interest. . .The more this power is exercised in the direction of promoting rapid circulation among the plates of which the great battery is composed, the greater is the tendency to the development of an inspiration and an energy closely resembling the service of the lightning of heaven subdued to human use.

In this passage Carey signals the advent of a new rhetoric, another form of the industrial Edenic, which we can term, following Leo Marx, the rhetoric of the electrical sublime. The passage itself indicates how Carey utilized the dialectical categories positive and negative, not as antithetical terms but as signifying a unity among opposites. Thus disharmony and conflict are mere appearances that point to underlying harmonies. Similarly, as a form of popular culture, the rhetoric of the electrical sublime attempted to merge all those contradictory desires of the American imagination. Electricity promised, so it seemed, the same freedom, decentralization, ecological harmony and democratic community that had hitherto been guaranteed but left undelivered by machanization. But electricity also promised the same power, productivity and economic expansion previously guaranteed and delivered by mechanical industrialization. Other events that occurred during the decade that Carey penned the preceding passages presaged which of these contradictory desires were to determine American social policy. During the 1870s Edison and Bell developed the electronic technology that was to be the basis of the new civilization; Gould, Vanderbilt and others carried on the "telegraph war" and other patent fights for the right to control the new technology; and the basis for industrial giants such as General Electric that were finally to exploit the new technology was perfected. Edison, Bell and other wizards were exploited as symbols of the new civilization, utilized to curry public favor and demonstrate the beneficence of the new technology, while new empires in communications and transportation were created behind the mask of an electrical mystique.

Not everyone was mystified about the real meaning of the new technology. Intellectuals, however, both in Europe and America, could devise nothing more effective than a purely literary strategy for dealing with the situation. Jacob Burkhardt and Anatole France in Europe, Henry Adams and Samuel Clemens in this country, devised the strategy of inverting the technological sublime and portraying the new technology as a specter of disaster. In his novel, *Connecticut Yankee in King Arthur's Court,* Samuel Clemens published what was probably America's first dystopia or anti-utopian science fiction. The American idea comes full circle in the novel as Hank Morgan is projected backward in space and time only to be encircled when he realizes that the electric fence erected by his own order for self-protection actually entraps him. This is an important event in American letters precisely because it contrasts so sharply with the Whitmanesque optimism of his earlier Mark Twain work. Similarly, Henry Adams was obsessed with the laws of thermodynamics and the specter of entropic disaster. *The Education of Henry Adams* is filled with pages of disillusionment of the sort that led Adams to locate the exact shift from the "old universe" of Boston genteel culture to the new phase of history determined by sheer power in the events of the year 1844: "the opening of the Boston and Albany Railroad; the appearance of the first Cunard steamers in the bay; and the telegraphic message which carried from Baltimore to Washington the news that Henry Clay and James K. Polk were nominated for the Presidency."

Despite the morbid views of literary intellectuals, the rhetoric of the electrical sublime was appropriated by reformers and regenerated by visionary utopians. The reformers and idealists blamed the corporation for defeating the possiblities of the electrical revolution. Edward Bellamy's speculations in *Looking Backward* and *Equality* and William Dean Howells' fictional *A Traveller from Altruria* were reversions to sublime aspiration and returns to optimistic attitudes toward electricity. Bellamy, a socialist propagandist, and Howells, a genteel reformer and member of the Boston Bellamy Club, envisioned the social use of radio and television and rapid transport. For them electric power for communication and transportation were to facilitate the diffusion of culture, dispersion of population and decentralization of control: in Howells' phrase, "getting the good of the city and the country out of the one and into the other."

*   *   *   *   *

During the Depression American electricians contended that the promises of electricity had been subverted by "vested interests," but that hydroelectric power and a new type of political orgnization would redeem the original message of Geddes. During the 1930s, the "giant power" crusade was renewed but now under the auspices of government rather than industry. A spokesman for the League for Industrial Democracy, Stuart Chase, put forward "A Vision in Kilowatts" in *Fortune* magazine in 1933:

> It [electric power] not only marches to ever greater quantitative output but it also transforms the entire economic structure as it goes. In its full development, electricity can yoke a whole continental economy into something like one unified machine, one organic whole. The parts may be small, flexible, located where you please, but with their central station connections. Electricity can give us universally high standards of living, new and amusing kinds of jobs, leisure, freedon and an end to drudgery, congestion, and problems of a steam civilization. It can bring back many of the mourned virtues of the handicraft age without the human toil and curse of impending scarcity that marked the age.

The New Deal seized upon the motif of a "New Power Age" for its creation of the Tennessee Valley Authority and the Rural Electrification Administration. President Franklin Delano Roosevelt and his advisers invested T.V.A. and R.E.A. with the role of models for a new America, an inspirational symbol around which to rally people to renew their confidence in America and its capacity for rehabilitation. Addressing the World Power Conference in 1936, F.D.R. proclaimed the New Deal ideal of a pragmatist's utopia:

> Now we have electric energy which can be and often is produced in places away from where fabrication of usable goods in carried on. But by habit we continue to carry this flexible energy in great blocks into the same great factories, and continue to carry on production there. Sheer inertia has caused us to neglect formulating a public policy that would promote opportunity for people to take advantage of the flexibility of electric energy; that would send it out wherever and whenever wanted at the lowest possible cost. We are continuing the forms of overcentralization of industry caused by the character-istics of the steam engine, long after we have had technically available a form of energy which should promote decentralization of industry.

Roosevelt concluded that our command over electrical energy could lead to an industrial and social revolution, that "it may already be under way without our perceiving it."

The Tennessee Valley Authority was intended to serve as a showcase for the positive linkage of electricity, decentralization, and citizen participation in reclamation of the landscape. T.V.A. was not intended merely to generate energy and produce fertilizer. In the words of the President, it was also to grant the Middle South an exemplary way of life: "a social experiment that is the first of its kind in the world, a corporation clothed with the power of government, but possessed of the flexibility and initiative of private enterprise," "a return to the spirit and vision of the pioneer," which "touches and gives life to all forms of human concern." "If we are successful here," Roosevelt concluded, "we can march on, step by step, in a like development of other great natural, territorial units. . .and distribution and diversification of industry."

David Lilienthal's dedicatory address to T.V.A. summarizes and recapitulates the rhetoric of the electrical sublime:

> This valley will be the first to enjoy to the full the fruits of this new age, the Age of Electricity. Those who have its blessings in abundance will come into a new kind of civilization. New standards of living, new and interesting kinds of jobs, totally new industrial processes, an end to drudgery, congestion, waste. . .such things are in store for us. For in this valley in another decade, electricity will hardly be reckoned in cost, so cheaply can your communities then supply it.

The T.V.A. idea acquired numerous foreign admirers. Probably the most ardent was Madame Keun, a French visitor, whose *A Foreigner Looks at TVA* grasped the salient motifs of the American imagination that underlined the New Deal approach. In her book, the T.V.A. appeared as a "happy balance between the Jeffersonian dream of the self-sufficient agricultural community and the mechanical advantages of the power age." The T.V.A. experience, she thought, showed that "in a capitalistic democracy. . .that imperishable quest of man for the millenium can be pursued by evolutionary adaptation."

At home and abroad the T.V.A. ideal was considered the original model from which other regions and countries might adapt a prime vehicle of social democracy. In 1944 Henry A. Wallace argued for many T.V.A's around the world under the rhetorical rubric of "Universal Electrification." Wallace suggested that a postwar expansion of the T.V.A. would constitute a powerful force for peace, link economic interests on a noncontroversial basis and obviate international tensions from the Danube to the Ganges. For, as Wallace put it, "valleys are much the same everywhere." After World War II Arthur Schlesinger, Jr., saw the T.V.A. as a weapon in the Cold War that, if properly employed, "might outbid all the social ruthlessness of the Communists for the support of the people of Asia."

We shall not here review the fate of T.V.A.; it is an ambiguous legacy at best. Certainly it has not proved to be a vast and catalytic social experiment. Rather than being a progressive force in the economy, it has identified itself with the electrical goods industries that have clustered around it, and the Authority has even been accused of corrosive strip-mining practices and of rate fixing during the Dixon-Yates controversy. Rather than being a harbinger of economic and political democracy for the valley, it has bureaucratized its interests and rhetoric and identified itself with the status quo. Rather than leading to a new social age, it has merely used electricity to elevate the traditionally narrow and socially wasteful standards of efficiency. Indeed, the entire American romance with dam building, fertilizer and electrical power—both in domestic and foreign policy— increasingly looks like a profound misadventure. Exported as programs of development to other nations, it has involved the United States in political misadventures that equate American democracy with American technology and has resulted in proposals such as the Mekong Delta Authority and the McNamara project to electrify the DMZ. Applied to climatic zones different from those of the United States and Europe, the dam building mania has produced economic and ecological disasters.

The T.V.A. experience demonstrates the folly of identifying technical projects with the creation of democratic community. As contemporary rhetoric is doing with electronics, T.V.A. rhetoric coupled ideas about electrical sublimity with attitudes concerning contact with nature and saw in the merger the automatic production of democracy. As a result, T.V.A. mesmerized liberals and prevented a serious evaluation of its failures; only Rexford Tugwell seems to have maintained the degree of detachment necessary to understand that "T.V.A. became more of an example of democracy in retreat than of democracy on the march."

It was Lewis Mumford who placed the T.V.A. and similar ventures into greater perspective. In *Technics and Civilization* and other works during the 1920s and early 1930s, he had blamed "the metropolitan milieu" and "the cult of paper money" for the postponement of the new order prophesied by Patrick Geddes, Kropotkin, and the "garden cities" planners. By mid-century, a disillusioned Lewis Mumford found new culprits in the pragmatic liberals of the type who established T.V.A.:

> The liberal's lack of a sense of history carries a special liability: it makes him identify all his values with the present. . .
>
> Like their counterparts, in the Soviet Union and China, our own leaders are now living in a one-dimensional world of the immediate present unable to remember the lessons of the past or to anticipate the probabilities of the future. . . Similarly, the TVA is as characteristic of our American economy as DuPont or General Motors.

*   *   *   *   *

The idea of an electrical utopia, as we attempted to show in the first part of this article, took root in Europe with the Enlightenment and the onset of industrialism. It began as a literary convention adopted as a cultural strategy: an attempt to explore what Raymond Williams has come to call "The Long Revolution." Transferred to this continent, the idea of an electrical utopia emerges as a literal expectation as well as literature. European ideas had made us particularly susceptible to the notion that the creation of America signaled a new departure in social history. America's redemption from the past was to come from "Nature" and a "Virgin Land," a new scene of human society filled with unique possibilities. Developed into a political and social theory under labels such as "the theory of propitious circumstances," the rhetoric of the technological sublime forecast that mechanization and industrialization would not produce the untoward consequences apparent in the European version of the Industrial Revolution. Instead America was to realize, through a marriage of nature and mechanics, an unprecedented solution to the problem of industrialization, a solution that would rejuvenate all immigrants who ventured into the new world and allow us to transcend the typical evils of industrial society.

In the last third of the nineteenth century, it became steadily apparent that mechanical utopias were merely verbal inventions and that America was not to be exempt from European history. Rather than abandon this rhetoric, however, many intellectuals merely transferred notions of technological bliss from mechanics to electronics, and from the 1870s to the 1970s there has flourished a genre we have termed the "rhetoric of the electrical sublime." This rhetoric has invested electricity with the aura of divine force and utopian gift and characterized it as the progenitor of a new era of social life, which somehow reverses the laws and lessons of past history. Despite changes in vocabulary, the idea of an electrical utopia possesses a common rhetorical tendency whenever it has appeared over the last century: it invests electricity with the capacity to produce automatically, on the one hand, power, productivity and prosperity and, on the other, peace, a new and satisfying form of human community and a harmonious accord with nature.

As utopian thought it has been innocent if naive. But, like all mythology in a politically conscious age, the idea of an electrical utopia can be and is exploited by established institutions that promise resolution of the psychic and social dislocations of industrialism and also the enhanced riches and leisure of modern society. American intellectuals, the main curators of this rhetoric, always proclaim the future in word and then desert the future in fact. Technology finally serves the very military and industrial policies it was supposed to prevent. At the root of the misconceptions about technology is the benign assumption that the benefits of technology are inherent in the machinery itself so that political strategies and institutional arrangements can be considered minor.

The most distinguished and notable American commentator on ecology and the history of technology has been Lewis Mumford. Yet for all the restraint of Mumford's prose, the identification of electricity as the great instrument of social balance is still present. In his works on technics and civilization Mumford claimed that electrical energy and communication would lead to a decentralization of power and the restoration of man to a life in touch with nature. He contrasted the filth, disorganization and desolation of mechanical civilization with the cleanliness and aesthetic grace of a "neotechnic" world. He saw in the imagery of the electrical environment symbols of a new type of society. The caption of an illustration in *Technics and Civilization* (1934) reflected this mood:

> Interior of a giant power station. The calmness, cleanliness, and order of the neotechnic environment. The same qualities prevail in the power station or the factory as in the kitchen or the bathroom of the individual dwelling. In any one of these places one could eat off the floor.

Throughout his early works Mumford reemphasized the old dreams of the American imagination and their possible realization through electricity. He proposed a new metaphor, the electric grid, as the substitute for the assembly line and neat rows of mechanical type.

> The principle of the electric grid must be applied to our schools, libraries, art galleries, theaters, medical services; each local station though producing power in its own right must be able to draw on power, on demand, from the whole system.

And he saw in the new technology a generous accord with the landscape and nature.

> Against all these wastes the neotechnic phase with its richer chemical and biological knowledge sets its face. It tends to replace the reckless mining habits of an earlier period with a thrifty and conservative use of the natural environment. Concretely, the conservation and utilization of scrap metals and scrap rubber and slag means tidying up of the landscape: the end of paleotechnic middens. Electricity itself aids in the transformation. The smoke pall of paleotechnic industry begins to lift: with electricity the clear sky and clean waters of the neotechnic phase come back again; the water that runs through the immaculate discs of the turbine, unlike the water filled with the washings of the coal seams or the refuse of the old chemical factories, is just as pure when it emerges. Hydro-electricity, moreover, gives rise to geotechnics: forest cover protection, stream control, the building of reservoirs and power dams.

Mumford himself was aware of the various political and economic forces that could betray the promise of neotechnic civilization. Yet decades later he kept his faith that utopia was just around the corner: "the new period of sun power and electric power is at hand. . .the age of balance and organic symmetry lies before us: the dialectic opposite of the age of specialism, division and disintegration."

At this moment when the dream of the electrical pastoral should again stand revealed as an empty promise—when Patrick Geddes' prediction of progression from "slum to semi-slum to super slum" has become manifest—Marshall McLuhan and a covey of supporters have expropriated the scholarship of Geddes, Mumford and the late Harold Innis and converted it into a new mythology that proffers the rhetoric of the electrical sublime as a total world view. In McLuhan's work all the past claims for mechanical and electrical technology are identified with the latest electronic gadgets—computers, simulation, television—and projected into every domain of life: we are to have a decentralized order outside of the centripetal pull of megalopolis, a restoration of community and identity through "re-tribalization" and contact with nature, and freedom from the powerhouse of history.

*   *   *   *   *

The bias of American politics and modern societies, . . . is one that places a premium on the control of space over vast distances and the utilization of space for commercial and power ends. This spatial bias is most pronounced in the efforts of the railroad, the space program and the communications system. In social planning the spatial bias is expressed in the destruction of old neighborhoods and villages that have an interlace of human association, historic continuity, and sense of place—factors that once gone are irreplaceable. The spatial bias represents an attempt to plan, ecologically and socially, for the control of space by technological devices. These plans invariably seize upon technology as both symbol and fact, and, consequently, America and other countries have successively hailed the canal, the bridge, the highway, the hydroelectric turbine and the computer as vehicles not only for communication and transportation but also for metaphysical voyages toward a realm of possibility. It is in this context that the myth of the powerhouse is implanted into the pathways of theory and practice and acted out upon the real environment.

In *Brooklyn Bridge: Fact and Symbol* Alan Trachtenberg has shown how poets like Whitman and Hart Crane confused the fact and symbol of technology, composing verse hymns to ideals, which they then read into the tangible presence of machines and artifacts. "Walt Whitman," Trachtenberg writes, "failed to consider that the engineer could not open a path to nature without first taking an attitude toward the land and toward society that would very likely postpone the metaphysical passage to India." Trachtenberg also observes that Hart Crane in "The Bridge" tried to write a mystical synthesis of America whereby poetry itself might absorb and transform the disagreeable facts of industrial life. Both Whitman and Crane, as Trachtenberg notes, expressed disappointment and disillusionment at the end. He points out that Crane's aim was "to overcome history, to abolish time, and the autonomy of events. . .to show that all meaningful events partake of an archetype," that is, "the quest for a new world." Trachtenberg concludes that the failure of the poetic version of the rhetoric of the technological sublime was due to concrete and cultural realities: "A detached symbol," he writes, "whatever its roots in the psyche of the poet, is helpless against the facts of tunnels, cinemas, and elevators." Crane himself finally concluded that the bridge has "no significance beyond an economical approach to shorter hours, quicker lunches, behaviorisms and toothpicks." The Brooklyn Bridge, on clear days, might still provide a panorama of the city, but it has not stemmed the urban crisis by either fact or symbol.

Recently, the Gateway Arch in St. Louis has been another example of an attempt to link by fact and symbol the East and the West and also the contradictory parts of the American dream. The arch was projected to commemorate the pioneers of the Westward migration with the inspirational figures of Thomas Jefferson and Lewis and Clark. At the final dedication ceremonies in 1968, Hubert Humphrey, then Vice President, credited the archway with the power to move those who merely came in contact with it. He called it "more than a symbol, a living memorial which evokes in those who stand before it" the determination "to wipe out every slum" and "to mend every flaw by living up to the arch."

Although Humphrey had been an unabashed advocate of "energy for abundance" through electrical power projects and intimately associated with the policies of the Great Society in the expansion of technological America, he made the effort to reconcile conservation with technics, and quality with quantity:

> Let the Gateway Arch stand, thus, as a symbol of America's determination to have beauty with utility, quality with quantity, and humanity with progress. . .
>
> . . .None will leave this site without a renewed sense of the elemental qualities of beauty. Whatever is shoddy, whatever is ugly, whatever is waste, whatever is false, will be measured and condemned by this new standard of excellence. . .
>
> . . .There is room in America for the highway and the wilderness. America is vital enough to earn the dollar and help the dream.

We can compare two Republican versions of the rhetoric, one that ushered in the century, another that opened the seventies. In September, 1901, President William B. McKinley attended the Pan-American Exposition in Buffalo and there spoke on the material and moral progress of America and the world, "the triumphs of art, science, education, and manufacture which the old has bequeathed to the new century." In assessing the technology arrayed at the exposition McKinley stated:

> Modern inventions have brought into close relations widely separated people and made them better acquainted. Although geographic and political divisions will continue to exist, distances have been effaced.

Following this invocation, McKinley suggested that a new age had been brought about when developments in communication and transportation would serve not the cause of war and adventure but commerce and peace:

> At the beginning of the nineteenth century there was not a mile of steam railroad on the globe. Now there are enough miles to make its circuit many times. Then there was not a line of electric telegraph; now we have a vast mileage traversing all lands and all seas. God and man have linked the nations together. No nation can longer be indifferent to any other. And as we are brought more and more in touch with each other, the less occasion is there for misunderstandings and the stronger the disposition, when we have differences to adjust in the court of arbitration, which is the noblest forum for the settlement of international disputes.

McKinley concluded that out of the exposition might come for all peoples and nations not only "greater commerce and trade" but also "mutual respect, confidence and friendship" which would "deepen and endure."

On the next day McKinley was shot by Leon Czolgosz, who not only was unmoved by the rhetoric but also was an anarchist embittered by the less inspiring aspects of industrial life. Moreover, the visions of commerce and peace projected by McKinley foundered on continued international conflict and economic instability. The expansion of American investment in the western hemisphere often precluded the Pan-American solidarity symbolized by the exposition, while the annexation of Spanish possessions involved America in international affairs ultimately as a protagonist in World War and Cold War.

McKinley's paean to a new century was echoed in Richard Nixon's recent State of the Union message which ushered in the seventies. Nixon adapted the old commonplaces to the newest concern, the environment, and created another frontier to be tamed by technology:

> The 1970s will be a time of new beginnings, a time of exploring both on the earth and in the heavens, a time of discovery. But the time has also come for emphasis on developing better ways of managing what we have and of completing what man's genius has begun but left unfinished.

In order to forestall suggestions that the problems of ecological reform would involve any limitation upon the American powerhouse, Nixon admitted that an argument is often advanced that there exists a fundamental contradiction between economic growth and the quality of life, "so that to have one we must forsake the other." He reassured the Americans that "vigorous economic growth" would provide "the means to enrich life itself and to enchance our planet as a place hospitable to man."

It was Nixon's predecessor, Lyndon B. Johnson, who in our times has best exemplified the rhetoric of the technological sublime in American political culture. His Administration provided a case study in the mistakes resulting from an attempt to substitute machinery and technique for political action in domestic affairs and diplomacy in foreign policy.

The figure of L.B.J. and the imagery of the Great Society were set against a background of conflicting attitudes toward past and future. In a 1966 documentary, "The Hill Country," Johnson portrayed himself as a man of rural electrification, reclamation of the land and Jeffersonian agrarianism. William Shannon concluded after seeing the documentary that the real L.B.J. embodied the antithesis of conservation, populism and rusticity. The cameras did not show the oil wells, high financiers, television holdings and political operators characteristic of Johnsonian Texas.

In his Ann Arbor convocation speech in May, 1964, Johnson announced that:

> The Great Society is a place where. . .the city of man serves not only the needs of the body and the demands of commerce, but the desire for beauty and the hunger for community. It is a place where man can renew contact with nature.

The trouble with the Great Society was as much in its formulations as in its flounderings, for it did not question the assumptions of a technological society but sought to pursue contradictory desires. The conquest of space, a recurrent motif, was not only accepted but accelerated without thought to cost or priority. The thought that community could be created, that men could renew contact with nature, simply by providing public housing and additional acres of recreation and parkland, was merely a reduction of ideas about "quality" and "style of life" to spatial arithmetic.

In 1966 Robert Kennedy, expressing dissatisfaction with both the New Frontier, which he helped to formulate, and the Great Society, attempted to define, however vaguely, a conception of democratic community beyond technology:

> . . .the destruction of the sense, and often the fact, of community, of human dialogue—the thousand invisible strands of common experience and purpose, affection and respect, which tie men to their fellows. It is expressed in such words as community, neighborhood, civic pride, friendship. It provides the life sustaining force of human warmth and security among others, and a sense of one's own human significance in the accepted association and companionship of others.
> . . .Nations or great cities are too huge to provide the values of community. Community demands a place where people can see and know each other, where children can play and adults work together and join in the pleasures and responsibilities of the place where they live. . . .Yet this is disappearing at a time when its sustaining strength is badly needed.

While the rhetoric of the electric revolution can be pierced by study of its historical development, it can even more clearly be punctured by the analysis of contemporary events. For the supposed trinity of sublimity in electricity, of decentralism and naturalism and harmony, is sharply exposed by exactly opposite trends in American industry and policy. Indeed, it is more centralism, pollution and conflict that are brought to the fore by the new technology and its applications. Electric power generators, television networks and computers systems are used in fact to augment and extend the spatial bias and controls of the powerhouse society.

In *The Computer Age and Its Potential for Management* (1965), Gilbert Burck of *Fortune* magazine disputed the prophets of decentralization in corporations and other institutions and demonstrated that the reverse process takes place under computerization of decision making:

> . . .just as the cable and wireless once brought farflung and quasiautonomous ambassadors and proconsuls under home-office control, so the computer is now radically altering the balance of advantage between centralization and decentralization. It organizes and processes information so swiftly that computerized information systems enable top management to know everything important that happens as soon as it happens in the largest and most dispersed organizations.

Far from decentralization, Burck concluded, computers and automation have "generated what appears to be nothing less than a pervasive recentralization." The recentralization has been brought about not only in corporations and industrial plants, but also in government.

In the Pentagon and the State Department, the computer and other technological instruments have undercut the conduct of traditional international activity. The United States has virtually abandoned an important interest in diplomacy and a trained consular corps to carry on foreign relations. The criteria of the diplomat, the definition of national interest and other permanent realities of history and geography, have been supplanted by the errant effort to program policies in data banks; also, extravagant technical aid projects and jet-stops by high officials on worldwide tours have been substituted for diplomatic action.

Short of the constant threat of "accidental" nuclear war by a radar misreading or a defect in missile guidance systems, there is no more obvious example of the errors reinforced by the mystique of the computer and other electronic machinery than the Vietnam conflict. In the Pentagon and the State Department, the technical approach, in complete contrast to the politics of diplomacy, could not perceive basic factors of ideology, nationalism, ethnic and regional differences—historical forces that could not be overcome simply by computer planning, electronic surveillance of hostile elements, and electrification of the Mekong basin and of rural areas. The "pacification" and "strategic hamlet" programs were evaluated by computers that could not simulate the entire Vietnamese experience. One Defense Department spokesman recently complained that there was an "inundation" of data that actually operated against sensible decisions.

Vietnam was also an attempt to test what General William Westmoreland called "the automation of war." The "electronic barrier" across the D.M.Z. projected by Robert McNamara to stop infiltration of troops did not succeed, and military officers complained that it tied U.S. forces down during a mobile, shifting war. Further, it was expected by Lyndon Johnson that his co-optation of a proposal to electrify the Mekong Valley like an Asian T.V.A. and to install "security lights" in villages under R.E.A. auspices would both win the sympathy of the Vietnamese masses and stop Vietcong night operations. Indeed, the rhetoric of Lyndon Johnson and his Administration on these issues reveals a complete mesmerization by the mystique of computers, hydroelectric turbines, and electronic eyes, which was followed into terrible and tragic commitments of disastrous proportions.

Another example of the projection of electricity and technology by Americans into foreign affairs was the proposal of Lewis Strauss and Dwight Eisenhower to deal with the Middle East crisis by installation of electric power and desalination plants. This suggestion was taken up by the Johnson Administration and various liberal Democrats in Congress on the notion that the drama of technology and the prospect of prosperity would overwhelm the differences between Isralelis, Palestinians and Arab states. Golda Meir and Israeli officials, however, were embarrassed by plans to inject this into the foreign aid appropriations because it would detract from Israel's military priorities. The Arab response is somewhat similar, since neither side thinks, as American policy-makers suppose, that nontechnological issues can be managed by technological devices. Ultimately, it is politics and not technology that must solve such crises.

The second illusion about the electric revolution is that it would be inoffensive to the environment, that there might even be a creative marriage of ecology and technology at some point. This illusion has been propagated in part by mistaken intellectual theorists, but it has been most exploited by the very industrial polluters who profit from the postponement of enforcement of antipollution regulations. The Edison Electric Institute enjoins Americans to "live better electrically" because they say, "total electric design offers new comfort, cleaniness, efficiency." Consolidated Edison announces that "nuclear power plants don't make smoke . . . produce electricity more economically . . ." for "around the clock power." General Telephone and Electric suggests: "We're making nights safer by making lights brighter." To ward off criticism of pollution and disruption, Consilidated Edison also changed its slogan from "dig we must" to "clean energy." And American Telephone and Telegraph projects a both ethereal and "folksy" image by urging us to "Get together with our communications consultant soon . . ." and by assuring us that "Communication is the beginning of understanding."

Yet the tremendous increases in public demand and usages created by advertising and engineering of the electric power and communications industries entail destruction of the landscape and pollution of the natural and human environment. In *Urban Choices,* Roger Starr observes that, "The most pervasive airborne waste with which the cities must contend is the waste produced by electric generation." But "the one option they lack is cutting down their total power requirement or even stabilizing it," since the engineering officers and advertisers "spend their days thinking up new devices to increase the urban dependence on electricity." The alternative is to generate power outside the population centers, but this involves the pollution of the countryside and its air and water, and also the use of remaining land to implant giant power lines. The main response of the utilities companies has been to construct "ornamental" power plants and lines, request more plant sites, and project solutions into future safety devices and perfected nuclear generators.

As scientist Sheldon Novick shows in *The Careless Atom,* however, there is not yet any assurance that a safe and peaceful atomic energy can be expected. Still, the Atomic Energy Commission and the electrical industry try to obtain new reactor plants despite the resistance of the public and of state officials and congressmen to these questionable installations. The various safety devices for more conventional energy sources, meanwhile, have proved less than perfect, especially owing to the vast increase in electrical power generation and demands. New methods and market demands are in fact constantly created without any regulation so that no stop is put to the spiral of power pollution.

Questions of personal safety and environmental aesthetics have also not been adequately answered. The radiation hazard in color television sets was rationalized by Mr. E. W. Henry of the Electronic Industries Association as an "isolated incident," despite the severely limited checking procedures of the federal health service inspectors and the industry itself. The *carte blanche* given to electricity in the American city has also created a "neon wilderness" (as Nelson Algren called it) of enormous signs for commerce at the cost of urban sensibilities. The neon sign business, of course, also involves demands for more and more power to serve the commercial calculus, but without regulation and intensive limitation.

Recently, Senator Edmund Muskie told a publisher's conference in Chicago:

> As always, men and women will lash out against the obvious threats to their health and well-being. They will attack nuclear power plants and oil refineries, paper mills and automobile factories, tanneries and steel mills. . .
> . . .Our technology has reached a point in its development where it is producing more kinds of things than we really want, more kinds of things than we really need, and more kinds of things than we can really live with; the time has come to face the realities of difficult choices.

Nonetheless, the solutions offered to environmental problems often contain the seeds of further problems. The proposals for the electric and also the steam car, for example, are flawed. The electric car will require the generation of more energy and therefore pollution to charge its batteries. It will also vastly reimburse the giant utilities companies for their expenditures, and reinforce the existing economic monopolies and their powers. The controversial steam engine automobile may involve fewer pollutants and safety problems, although the rhetoric of its proponents sounds very much like that of the celebrators of the locomotive and steamboat of the last century. But it still would not solve the problem of excess traffic implicit in the American romance with cars.

The least dramatic, most serious problem posed by electric and electronic technology of communications and transportation, however, is its erosion of organic cultures and cultural institutions. The space-binding bias of the television networks, to a very large extent, is to disregard in tone and coverage local and regional developments and to ignore the cultural diversity of ethnic groups and nationalities. This creates what might be called the dilemma of the many silent minorities in America, particularly the key role and legitimate interests of the

European ethnic blocs in the urban crisis. Moreover, the interests of the electronic communications complex preclude any support by it for certain significant cultural and political changes.

In terms of communications policies, the United States is now committed to an enormous expansion with great international ramifications and repercussions. The rhetorical component of these policies is expressed in the various documents published by the Carnegie Commission on public television, the Ford Foundation on broadcasting, and the United States review panel on communications policies. Invariably, these papers reflect a spatial bias toward the focus and scope of communications media and an elitist bias of cultural *noblesse oblige* toward the receivers and subjects of the media within the United States. They also reflect a drive for American hegemony over the growth of world communications. At the present time, as Professor Harry Skornia has pointed out in *Television and Society,* the electronic means of communication are controlled by and concentrated in corporations, and this in turn reflects the dominance of the industrial powerhouse over the rest of American life. The media function as implicit instruments of socialization and social contol forming patterns of consumer behavior and an audience of passive spectators. The language of TV as a public service suggests that cultural needs and deficiencies will be compensated for by more programming when in fact that programming primarily transmits the bias of the corporations and foundations that finance and direct it.

Professor Herbert Schiller has recently documented, in his book on communications and American empire-building abroad, that the content and equipment for broadcast systems in new nations is almost entirely owned, leased or sold by U.S. interests. The Rostow communications review panel indicated that the government might go even further than the corporations and foundations by sponsorship of a single new monopoly under some form of government supervision to handle telegraph, telephone, radio and television. The panel, whose chairman was Eugene V. Rostow, then Undersecretary of State for Political Affairs, delivered a report to Lyndon Johnson that mirrored the common American cult of communications-for-communications'-sake. Referring to Comsat, the communications satellite, Chairman Rostow stated: "It holds no ideological goal except that it is good to communicate efficiently. . .seeks no diplomatic advantage. . . Here is a rare opportunity to join in an activity to bring benefits to all nations and loss to none. . . Nothing could better symbolize the truth that space belongs to all men, than an international undertaking that permits the free flow of communications. . ."

The rhetoric of electronic open-endedness, however, closes on a frankly political key: "Our nation is in a relatively poor bargaining position on communications matters with foreign counterparts since we do not speak with a single voice. Defense communications in the future could be subjected to delay. Are we making the best use of the electromagnetic spectrum?"

There are, of course, dangerous political implications in the review panel proposal with its formulation of a new megamonopoly. More fundamental American attitudes are reflected in the belief that technically efficient communications are good, that space has social symbolism, that conflicts of ideology and interest can be surmounted by technology. The larger significance of electronic technology of communications nontheless resides in its capacity for explicit ends, such as commerce and military use and powerhouse purposes.

In *The Accidental Century*, Michael Harrington has assessed the overall implications of the new communications media in its effects upon the dream of a social democracy in America and Europe as well: "In such a massified society, the countervailing capacity of the voluntary organizations has been either diminished or reduced to zero. . . Mass communications played an important role in the process. They tended to standardize reactions to root out the particular world view of the neighborhood or the plant and to impose a tame, 'nationalized' way of thinking and feeling upon the people."

The American romance with technology, and the attitudes and behaviors it nurtures, has also been summed up by the historian William Appleman Williams: "America's great evasion lies in the manipulation of nature to avoid a confrontation with the human condition and with the challenge of building a true community."

Lewis Mumford, in his moments of disillusion with the idea of utopia, has also recorded the mirages and evasions associated with the myth of the powerhouse. Mumford has suggested that Americans felt that the "problem of justly distributing goods could be side-tracked by creating an abundance of them. . .that most of the difficulties that had hitherto vexed mankind had a mathematical or mechanical. . .solution." Beyond quantitative solutions we need, Mumford has written, "a conception of what constitutes a valid life. . ." What has to be challenged, he observes, "is an economy that is based not on organic needs, historic experience, human aptitudes, ecological complexity, but upon a system of empty abstractions" such as power, mobility and growth. The very value system of the powerhouse society, Mumford submits, has to be undermined: "The problem of quantity, the problem of automatism, the problem of limitless power, which our very success in perfecting machines has raised, cannot be solved in terms of the machine."

Around the world, however, there has been a pronounced trend to place science in the role of secular religion and to accept technological determinism. In *The Failure of Technology*, Friedrich Juenger wrote that technology has become the *real-metaphysik* of the twentieth century. Juenger noticed that the perfection of technology involved the exhaustion of nature and natural resources and the reduction of man and the human scale. The perfection of technology would be in contradiction to human and natural survival. The predicted "finale" of technical evolution involves a contradiction and is necessarily "utopian because [it proposes to] combine what cannot be combined."

Such intellectual warnings have proved to be no competition to the "rhetoric of the technological sublime." Contemporary popular culture and the contemporary heroes of youth exude electrical metaphysics as the new religion. John Lennon, doyen of the "hip" generation, speaks directly to the faith:

> Yes, I believe that God is like a powerhouse—the place where you keep electricity, like a power station. And that he's a supreme power, and that he's neither good nor bad, nor left, right, white or black. He just is. And I believe that we tap that source of power and make of it what we will. As electricity can either kill people in a chair or light a room. I think God is.

At the root of much of this mythic thinking about technology resides an essentially primitivist argument. In reviving the ancient drama of the "dialectic of the middle way," modern technophiles have allowed an interesting twist of some anthropological importance to creep into the script. The scene or setting for the drama of the mechanical sublime was Nature into which the machine was absorbed and humanized. The setting of the modern drama is now technology itself, which has absorbed nature and, mysteriously, taken on all those qualities formerly attributed to nature. The new rhetoric is bathed in electrical images of communion and association, peace and serenity, integration and wholeness. As sociologist Paul Meadows put it: "Electricity [provides] a new technological base. . .which promises to make possible a personalized society, in which whole human beings participate as significant persons, not simply as specialized functionaries, in the situations of collective living." Now it is the person, the human self, who is the outsider, the intruder, who is to be restored by being absorbed into technology.

The appeal of this electrical drama represents a surfacing of a deep mythopoetic prompting of the self, albeit a prompting with many political uses. Ernst Cassirer, the late German philosopher, observed that man's discovery of himself as a symbol-making creature, alien to the natural world, was a tragic discovery, a discovery of human isolation. To heal the breach between man and nature, primitive man identified his life with that of nature and particularly its animal inhabitants. As Joseph Campbell has suggested, the problem of primitive societies tied closely to nature was to become linked psychologically with primitive forces. Through acts of imitation, totem worship, ritual and mime, persons identified the life rituals of birth and death with the rhythmic and seasonal movements of nature. They thereby humanized the scene of existence and brought foreign forces under spiritual control.

The environment of modern man is not nature but technology. Still the same desire and necessity to be linked symbolically to the environment is present. This linkage is forged by attaching biological rhythms to industrial ones, by modern forms of totemism and by producing equations between man, nature and technics. In *Attitudes Toward History* Kenneth Burke observed that the tone of awe that permeated the discussions of electricity following the New York power failure of 1965 led him to believe that "if some new cosmic radiation entered the solar system to make the performance of electricity a bit irregular as with the weather, urban populations would start praying for electricity as farmers pray for rain."

The identification goes deeper than that, however. From the onset of industrialism there has been a persistent attempt to create mechanical metaphors to describe the processes of bodily action and the mental process and thereby to develop a universal principle to explain all reality, human and nonhuman. When electricity replaced mechanics as the basis of technology, the principles of the new technology, as we have attempted to show, were adduced as the underlying "laws" of all reality—nature, man and society. Moreover, electricity particularly lent itself to this universal identification for, unlike mechanical power, it is a mysterious force, unseen, which apparently resides in the very atmosphere, and leaps into action at a touch. Electricity has been regularly seen as a divine force, expressing the secret of life, somehow transcendent, not artificial, and related to the sun and solar power. It is therefore seen as natural and social, ancient and modern, and as providing a unique and providential bond between man and his environment.

McLuhan's prose, in particular, invokes this primitive mythology. In *Understanding Media* and his more recent work, man enters a symbiotic relationship with technology; he becomes an analogue of the machine, possessed and controlled, invested with the properties of computers and electronics. His body is characterized as an extension of technology, and his mind as a complex, electrically charged computer. By such verbal alchemy men are identified with technology, and the breach between man and machines is symbolically healed.

We are suggesting that there is a deep, primitive mythology invoked in the appeal of McLuhan and others of his school: the universal desire of the self to be organically connected to his environment. This connection is always symbolic rather than empiric. With the substitution of technology for nature as the actual environment of man, the primitive dialectic is reenacted but now in a process by which electrical technology is humanized and man electrified. As McLuhan has said: "The new media are not bridges between man and nature; they are nature."

While the identification of man and his environment is always a principal task of mythology, what one must question are the political uses and psychic effects of such identifications. Ironically, McLuhan once understood this. In *The Mechanical Bride* he quoted the lines from Joseph Campbell we have just paraphrased and followed that quotation with these comments:

> It is the same annihilation of the human ego that we are witnessing today. Only, whereas men in those ages of terror got into animal strait jackets, we are unconsciously doing the same *vis à vis* the machine. And our ads and entertainment provide insight into the

totem images we are daily contriving in order to express this process. But technology is an abstract tyrant that carries its ravages into deeper recesses of the psyche than did the sabertooth tiger or the grizzly bear. . . As terrified men once got ritually and psychologically into animal skins, so we already have gone far to assume and to propagate the behavior mechanisms of the machines that frighten and overpower us.

McLuhan has since recanted on his own observations in order to support the "new commonplaces" of corporate America. The identification of electricity, mind and divinity, however, knows no particular political or intellectual affiliation. Maxwell Maltz uses the same electrical metaphors to promote "psycho-cybernetics" and restore the flagging optimism of the traditional middle classes. Maltz promotes electrical language as the key to success and happiness and insight into the universal mind. It is but a step from Maltz to McLuhan's belief that through electronics we will bypass not only literacy but language itself and approach a state of speechlessness, wherein by mind reading and direct participation in the totality of knowledge we shall enter a "perpetuity of collective harmony and peace." But the linkage of electricity, extrasensory perception, mind reading and instant unmediated communication is a standard offering in the tradition of utopia. In the nineteenth century, Edward Bellamy developed the notion of thought reading in a short story entitled prophetically "To Whom This May Come." The story describes a lost race of mind readers and their "power of direct mind-to-mind vision, whereby pictures of the total mental states were communicated instead of the imperfect descriptions of single thoughts. . ." Bellamy developed the utopian possibilities of mind reading for a "Religion of Solidarity":

> But think what health and soundness there must be for souls among a people who see in every face a conscience which, unlike their own, they cannot sophisticate, who confess one another with a glance and shrive with a smile! Ah friends, let me now predict, though ages may elapse before the slow event shall justify me, that in no way will the mutual vision of minds, when at last it shall be perfected, so enhance the blessedness of mankind as by rending the veil of self, and leaving no spot of darkness in the mind for lies to hide in. Then shall the soul no longer be a coal smoking among ashes but a star set in a crystal sphere. . .

In *Looking Backward,* Bellamy contemplated a revolution in the human condition by an intensification of psychic awareness. He described the change as an "indistinct revolution more radical than if it had been political." Using the metaphor of electricity, he describes a state of cosmic awareness, a bypassing of communication, a loss of self and direct unmediated participation in the totality of life and knowledge.

The phenomenon to which Bellamy refers we would call today, following Teilhard de Chardin or Julian Huxley, the noosystem or noosphere: the belt of organized intelligence that encloses the planet and the biosphere. For Huxley this is "a single pool of common thought greatly helped by modern means of communication [which] is putting an end to parochial inhibitions." From this Buckminster Fuller believes that there will emerge a "dynamic, electronically articulate, constant world democracy referendum whose computer integrated voice and evolutionary wisdom will be stunning. . ."

Because electricity propagates visible and audible messages by invisible means, there is a persistent attempt to link it with an extrasensory and wordless world that can be entered in an unmediated way. There is also the parallel belief that electrically propagated knowledge floats above the biosphere on gossamer wings, providing a curtain of heaven for the human scene.

Such dreams merely confirm the consistent American strategy for responding to industrial realities and reveal again the contradictory dreams of American intellectuals. The American character since early in its history has been pulled in two directions and has been unable to commit itself to either. The first direction is toward the dream of the American sublime, to a virgin land and a life of peace, serenity and community. The second direction is the Faustian and rapacious, the desire for power, wealth, productivity and universal knowledge, the urge to dominate nature and remake the world. In many ways the American tragedy is that we want both these things and never seem to respect the contradiction between them. The dream of the American sublime has never been used to block the dream of American power primarily because the rhetoric of the technological sublime has collapsed the distinctions between these two directions.

As a result, America has allowed technology and the urge for power unrestricted room for expansion. Americans have also tried to insulate themselves from technology's more destructive consequences by projecting a zone, a spatial place, outside of and independent of the destructiveness of industrial society. Nature, the first such zone, was naïvely seen as impervious to the force of machine technology. It remained the middle landscape, the zone of peace and harmony. When nature fell before the powerhouse center of society, the zones were made metaphysical: new tribes, world communities, omega points, cosmic states of consciousness.

The theme of the electronic revolution today continues ancient attitudes of American popular culture and serves up contradictory messages to the American public. For American communications and electrical goods industries, the rhetoric of the electronic revolution provides a rationalization of the *status quo* and a legitimation of the purposes of the new technology. For the mass public, the rhetoric resuscitates those themes of a sublimity that contain the contradictory desires of the American imagination.

In the early stages of industrialization, American popular culture looked back at the receding landscape that was celebrated in popular science, art and literature, the preindustrial world, the world of nature and the folk. Now—with nature unknown, the landscape lost, family displaced, with history a mere echo—modern man, still unable to endure the present, turns to the future, away from the world that was to the world that might be. As science fiction replaces—in literature, story, song and cartoon—the world of the frontier, the pastoral haven, the pristine village, so too the popular intellectual marketplace replaces odes to the world of Nature with paeans to the world of Technology. In this Disneyland of the mind, the same audience queues up, but the main attraction is no longer Frontierland but Tomorrowland. McLuhan's exploitation of this postindustrial metaphor does not signal the rejuvenation of American social thought but, given our present difficulties, its exhaustion.

What, then, is the responsibility of intellectuals concerning the electronic revolution?

We submit that it is not the convocation of a Vision '80, a Mankind 2000 project or a congress of futurologists. The history of the theory of "neo-technic utopia" reveals that an intellectual involvement in elaboration of plans for the application of technology has been an inadequate approach. These attempts have failed because the bias of electric power and communication is antithetical to dispersed use and small-scale control.

The promotion of the illusion of an "electronic revolution" borders upon complicity by intellectuals in the myth-making of the electrical complex itself. The celebration of the electronic revolution is a process whereby the world of scholarship contributes to the cults of engineering, mobility and fashion at the expense of roots, tradition and political organization. As Harold Innis pointed out, the demise of culture could only be offset by deliberately reducing the influence of modern technics and by cultivation of the realms of art, ethics and politics. This requires action to counter and direct rather than disguise the bias of the electronic revolution; it means cultural and qualitative checks rather than more quantitative definitions of the quality of life; it requires defusing the humanistic from the technological instead of offering a contradictory image of humanized technology.

Obviously, the electronic revolution cannot be managed by purely literary strategies, by creating images of the antisublime, or "black humor" allegory, or by creating new zones of romantic isolation and innocence. Yet this is precisely the profession of Orwellian science fiction and confrontation-protest techniques. These are neo-luddite activities that bespeak a belief that apocalypse is upon us and that only a symbolic crusade, "wounding the Pentagon," or exorcising bad Karma, can save us. Like that electronic revolutionaries, antitechnologians suggest that we are living in a new age unlimited by previous history, politics and technology. They merely reverse the mythology about the electronics power-houses. In a faulty response, they seek illusory mirages, reprogrammed sensibility, a chemical pastoral, a politics of style. By bypassing the steady work of scholarship and politics, this engagement in intransigence, resistance and electric circuses only Americanizes the "myth of Sisyphus." Paranoia about mass media and a sense of powerlessness are the simplistic obverse of the *mythos* of the electronics. The stance of powerlessness debilitates and means more powerlessness.

In *Player Piano,* Kurt Vonnegut predicted the ultimate defeat of any neo-ludditry in his "America in the Electronic Age." As his would-be counterrevolutionaries proclaim: "Those who live by electronics die by electronics." "We'll rediscover the two greatest wonders of the world, the human mind and hand. . .walk wherever we're going. . . And read books, instead of watching television. . ." But the new ludditry fails to offer alternative ways of life. Consequently, the technological imperative expands into new domains despite the protest.

There is another *zeitgeist* of irrelevance that tries to set up special locations and insulations, literal human reservations, where electronics can be mastered and tamed by secular prayer and imagination. In B. F. Skinner's *Walden II* and Huxley's *Island* a merger of folklore and futurism takes place on the artificial plane of utopianism. In Huxley's words, "electricity minus birth control plus heavy industry equals totalitarianism, war and scarcity," but "electricity plus birth contol, minus heavy industry equals democracy, peace and plenty." This is the type of *new* commonplace that renames manipulation as rehabilitation, technocratism as humanism, *et cetera.*

We advocate that intellectuals deal with realities and speak to the living concerns of the populace rather than escape from politics or return to folklore. At present, it is incumbent upon us all to resuscitate what remains of a universe of discourse, political language and democratic vocabulary. Already, our conceptual and perceptual capability has been bombarded and our moral dimensions denuded by the mythology of technology and the folklore of a past idyllic.

The first task is to demythologize the rhetoric of the electronic sublime. Electronics is neither the arrival of apocalypse nor the dispensation of grace. Technology is technology; it is a means for communication and transportation over space, and nothing more. As we demythologize, we might also begin to dismantle the fetishes of communication for the sake of communication, and decentralization and participation without reference to content or context. Citizens now suffer in many areas from overloads of communication and overdoses of participation. We should address ourselves directly to the overriding problems: the uprooting of people from meaningful communities, and the failure to organize politically around authentic issues. Thus, functional participation and geographical decentralization cannot solve problems in government, factories and schools that are constitutional and not merely mechanical. The political questions are not centralism versus decentralism but democratization; not book versus computer in education but an adequate curriculum; not representative versus participational institutions but the reconciliation of immense power and wealth with the ideals of liberty and equality. It might be that real control over electronic media will necessitate more formal centralism. The point is really a pragmatic one in the nonphilosophical sense of the word.

To reduce the twin delusions of technics and myth, we must convey these concerns to the public. Intellectuals should demonstrate the relevance of scholarly integrity and rationality by critical studies that can reach an audience in sympathetic terms. The focus should not be negativistic but in favor of the values of the arts, ethics and politics where man finds fulfillment. As Perry Miller put the matter in his eloquent essay on "The Duty of Mind in a Machine Civilization":

> "We may say without recourse to romantic isolationism that we are able to resist the paralyzing effects upon the intellect of the looming nihilism of what was formerly the scientific promise of bliss. . . millions of Americans, more than enough to win an election, have only vague notions, barely restive worries, as to the existence of any such enmity. . . Upon all of us, whoever we may be, rests the responsibility of securing a hearing from the audience. . ."

That hearing must be secured in a language of democracy that is demythologized and in which political words are again joined to political objects and processes. At least this seems to be a responsibility for formation of "a party of the mind."

In many ways, Mr. Herman Kahn seems to embody the more colorful aspects of the "Technetronic Era" in his own career. After serving as physicist and military analyst for RAND, he helped found and became director of his own think-tank, the Hudson Institute. After publishing such highly controversial strategic studies as *Thinking About the Unthinkable* and *On Escalation,* he undertook the role of "futurologist." Here, with his colleague Anthony Wiener, Mr. Kahn develops some fascinating speculations about the nature of American society in the year 2000. Making frequent use of historical and cross-cultural comparisons, the authors consider how styles of life and standards of behavior may change under the pressure of accelerating technology. Most significantly, Kahn and Wiener touch upon a recurring theme in analyses of the postindustrial revolution: the complex interaction of affluence and alienation.

Herman Kahn and Anthony J. Wiener on the Year 2000 •

A. INTRODUCTION

The purpose of this chapter is to continue outlining our Standard World, focusing here on domestic issues for the United States and related socio-economic changes that may also impinge on other advanced societies.

In Chapter I we pointed to a basic, long-term multifold trend which we judged likely to continue during the last third of this century (see Table I, Chapter I). Some of the elements of the trend—especially increasingly Sensate (empirical, this-worldly, secular, humanistic, pragmatic, utilitarian, contractual, epicurean or hedonistic, and the like) cultures; accumulation of scientific and technological knowledge; institutionalization of change, especially research, development, innovation, and diffusion; increasing affluence and (recently) leisure; urbanization and (soon) the growth of megalopolises; decreasing importance of primary (and recently) secondary occupations; and literacy and

•Source: Reprinted with permission of the Macmillan Company from *THE YEAR 2000: A Framework for Speculation on the Next Thirty-Three Years,* pp. 185-220 by Herman Kahn & Anthony J. Wiener. Copyright © by The Hudson Institute, Inc.

education—should become especially prominent in the "Visibly Postindustrial" countries projected in Table XXII (Chapter I). These items are likely to become especially conspicuous in the United States, followed perhaps rather closely by Canada, Switzerland, Scandinavia, and, with some cultural differences, Japan. The situation in the rest of Western Europe and the "Early Postindustrial" countries listed on page 60 will probably differ only in degree. Even the projected "Mass Consumption" and "Mature Industrial" economic groups listed on that page are likely to be quite different from their current early industrial conditions, and will probably begin to show "Postindustrial" features in leading sectors of their economies and societies, partly as a result of their own development and partly because of the dominant influence of the most affluent nations both on the world economy and on the styles of life of cosmopolitan elites everywhere. Indeed, even the remaining nations—more than half the world's population, with comparatively rudimentary industrialization and per capita incomes for the most part still below six hundred 1965 dollars—will show the influence, among privileged groups, and in major cities, of postindustrial changes.

B. THE POSTINDUSTRIAL SOCIETY

We introduced Daniel Bell's term "postindustrial society" in Chapter I, and listed fifteen items, repeated in Table I, below, that seem to us likely to be

TABLE I.   The postindustrial (or post-mass consumption) society

1. Per capita income about fifty times the preindustrial
2. Most "economic" activities are tertiary and quaternary (service-oriented) rather than primary or secondary (production-oriented)
3. Business firms no longer the major source of innovation
4. There may be more "consentives" (vs. "marketive")
5. Effective floor on income and welfare
6. "Efficiency" no longer primary
7. Market plays diminished role compared to public sector and "social accounts"
8. Widespread "cybernation"
9. "Small world"
10. Typical "doubling time" between three and thirty years
11. Learning society
12. Rapid improvement in educational institutions and techniques
13. Erosion (in middle class) of work-oriented, achievement-oriented, advancement-oriented values
14. Erosion of "national interest" values?
15. Sensate, secular, humanist, perhaps self-indulgent criteria become central

characteristic of the emerging situation, especially in the United States. (Some of these items are also part of the multifold trend and were discussed in part D of Chapter I, pp.34 ff., expecially pp. 57-64.)

We shall discuss most of these issues by simply mentioning them, or occasionally conjecturing very briefly on their significance, without attempting to be systematic or exhaustive. Many of the statements will nonetheless be in conclusionary form, mainly because it is easier to focus or provoke discussion in a short paragraph if one states a position than if one attempts to give several alternatives in truncated from. In all cases we tend to "believe in" the position indicated, though perhaps tentatively. These conjectures are no more than our current working hypotheses and are clearly more controversial than most aspects of our Standard World.

It may be useful to compare the shift away from primary and secondary occupations to an earlier transition made in the United States in the mid-nineteenth century. In that predominantly agricultural society, 55 per cent of the labor force were farmers, while today only about 5 per cent of the labor force is engaged in agriculture. One might imagine an average American in 1850, given the information that in one hundred years only 5 per cent of Americans would be engaged in agriculture: he would be very likely to think of these farmers as likely to use their "monopoly powers" to get far more than a proportionate share of power, influence, and resources. He might even assume that the power, prestige, and influence, then held in his day by the 55 per cent of the American people who farmed, would be held by the 5 per cent; or at least that the 5 per cent of the American population who had a monopoly on food production, the most basic of all activities, would be able to exploit their commanding position to overwhelming effect. An American who subscribed to the then widely held Jeffersonian idea that only the freeholder, as opposed to the propertyless laborer, could really participate responsibly in a democracy would have been especially likely to assume that the farmers, or at least the landed gentry, would continue as the most important social group.

While today many believe that farmers have more than their share of government subsidies and political power, especially in unreapportioned legislatures, all would concede that farmers are, by a large margin, not the dominating force in American politics. Furthermore, while continued improvements in agricultural productivity are of great importance, current productivity is so great that so far as the average American is concerned, improvements are routine and of little interest. In the eighteenth or even the nineteenth century, a technical advance in agriculture would have been noted with enthusiasm by the whole community and might have transformed the whole community; today only farmers are likely to be very interested in agricultural innovations.

A similar transformation, though not so extreme, may develop with respect to business activities in a postindustrial society. A smaller percentage of people may be engaged in business, and the very success of private business may make its further successes seem less exciting and dramatic. While businessmen will probably continue to be deeply occupied with their affairs, the issues of finance, investment, production, sales, and distribution that have so long been dominant concerns of so many Americans and Europeans will very likely dwindle in interest. American industry has already been concerned about its declining attractiveness for college graduates, especially for the most intellectually gifted segments of the group, and there may eventually be a general lowering of business morale.

The postindustrial era is likely to be more of a "learning society" than today's. In part this is because of the "information explosion," but mostly because of the rapidity of change. As we mentioned in Chapter II, the computers of 1967 have about ten times the performance capacity of those of 1964 and 1965, which means that concepts of appropriate computer functions that were perfectly valid for the computers of two or three years ago must be reviewed and sometimes completely changed. In many cases the new concepts that must be devised in order to deal adequately with today's potentialities are very different from those of two or three years ago.

While the computer is a fairly extreme example of rapid change, it is reasonably indicative—though the more likely time for significant change in most areas would be closer to ten or twenty than two or three years. Thus, if the annual per capita increase in income is 4 per cent, then per capita income will double every eighteen years. Such a doubling is clearly a most significant event. Or to take another example, television, which was almost unheard of in private homes in 1946 and 1947, had by 1955 changed American living patterns in a very marked way. In this case, the exact significance of the changes is yet unevaluated. (For example, it is difficult to be sure whether or not it makes a great difference, or how, that large numbers of children and their parents spend an average of several hours every day passively watching a flickering screen.) To give another example, it takes about twenty to fifty years for most countries to double their populations, so there may easily be two or three doublings within one lifetime. Similarly, a country may within one lifetime, change from being largely or overwhelmingly rural to largely or overwhelmingly urban. Or the number of kilowatts generated per year, the passenger miles in autos or planes, numbers of telephones, and so on may increase by factors of five to a hundred in a decade or two. Any of these changes occurring by themselves could be important. If all of them, and others as well, occur more or less simultaneously, the total rate of change and the need for large adaptions become fantastic. The term future shock has been used to describe the corresponding "acculturation" trauma. These changes seem likely to increase in number, kind, and rate at least for the next three or four decades. The corresponding need for adjustment, adaptation, and control is likely to be one of the most characteristic and central phenomena of the early postindustrial era, although there is some possibility that it will not be either as pervasive or dramatic in a later time period, simply because there may be a deliberate slowing down in order to "take it easier" and to avoid or meliorate consequences of change.

C. CROSS-CULTURAL COMPARISONS

We pointed out in Chapter I that in the previous six thousand years of recorded history civilized man, by and large, lived in societies not too dissimilar economically from that of Indonesia today, in which the average income was, give or take a factor of two, the equivalent of about one hundred dollars per capita. (Even the size of the larger societies, e.g., the Roman Empire or the Han Empire, was about the same in population as Indonesia today-about one hundred million people.)

The Industrial Revolution was, in many ways, a more important and more rapid change than any that had preceded. The changes that seem likely in the next thirty-three years also seem likely to lead to some results that are entirely unprecedented. Nevertheless we should suggest that some of the prospects for the year 2000 are, in effect, a return to a sort of new Augustinian age. Conditions (in the superdeveloped countries at least) could then differ from those of the early and mid-twentieth centuries—in some important ways—much as the early Roman Empire differed from the preclassical world. We are all too familiar with cliches about the decline of the Roman Empire, but for the better part of the first two hundred years the Roman Empire enjoyed almost unparalleled good government, peace and prosperity, and it should be noted that it also began in an "age of anxiety" and apprehension. Arguments are often heard to the effect that the "moral fibre" of the Romans somehow "degenerated" because of a lack of challenge during the period of stability and prosperity. While the issues of cause and effect are complicated and inherently inconclusive, there are some parallels between Roman times and ours. Various analogies, however trite or inaccurate in their usual formulations, ought not be dismissed without some thought, at least as sources for conjectures.

Thus it is interesting to note that when Augustus came to power the free citizens of Rome had 76 holidays a year. When Nero died, not quite a century later, they had 176 holidays a year. In our world, if productivity per hour goes up by 3 or 4 per cent a year (or by a factor of three or four by the year 2000), it is not likely that all of this increased productivity will be used to produce more. As in the Roman Empire, much may be taken up in increased leisure. One can almost imagine that in the next thirty-three years the once populist, bourgeois, conforming Americans become relatively elitist, antibourgeois, and pluralist;

significant numbers acquiring attributes that bear comparison to sophists, epicureans, cynics, primitive—or humanist—sensualists, other materialists, and various kinds of dropouts—to mix terminology usually applied to the Hellenistic era with current jargon. As in the Hellenistic and Roman Empires, such groups, while reasonably identifiable, would not necessarily be exclusive or distinct and many individuals might move more or less freely among the various groups. Nevertheless an observer of society might justifiably generalize about the existence of such groups—certainly more justifiably than he could have in traditional America.

What is meant by sensualists, materialists, and dropouts is probably sufficiently clear. Sophists have had a bad press; they are often described as unscrupulous and/or superficial manipulators of words and ideas. While the Greek sophists may have included some like this, very likely most resembled, at least in their philosophic ideas, modern pragmatic positivists or nominalists. The Greek cynics, stoics, and epicureans were all antiwar, indeed, rejected most worldly ambitions and pretensions to an almost escapist degree—in part as a result of their disappointment at what had happened to Greek culture after it had conquered most of the world. The cynics (of whom Diogenes is the best known) went furthest in this direction, scorning established values and aspirations, repudiating accepted codes of behavior, and arguing for a return to poverty and personal independence, and to nature (like "dogs," as the etymology suggests). Unlike the current "hippies," however, they cultivated self-control and asceticism. Most interesting of all to us are the stoic and epicurean reactions—both were basically pessimistic about the "meaning and purpose" of man's efforts (Gilbert Murray's "failure of nerve"), but were still, in their separate ways, philosophies of good and respectable citizens. The stoic, of course, had an almost biblical passion for righteousness and duty, while the epicurean had an almost similar passion "to cultivate his garden and friends." Tables II and III below give capsulated descriptions of these two important reactions.

TABLE II. The (Roman) stoic outlook

1. A passion for virtue and for doing one's duty
2. The four cardinal virtues are wisdom, justice, courage, and temperance
3. But fortitude and indifference to pain and sorrow are also important
4. Only such virtue justifies power
5. But virtue is its own reward
6. Even a slave can be virtuous
7. There is a basic natural law governing mankind
8. Under this natural law all men are basically equal

TABLE III. The epicurean outlook

1. Based on emphasizing sensation, emotions, pleasure of the individual soul
2. Criterion of good and evil is sensation, i.e., "pleasure"
3. Power and public life disturb the soul
4. When the body dies, it returns to atoms; when the soul leaves, the body has no "sensations"
5. There should be no fear of death, gods, or natural laws
6. Pleasure is "absence of pain," not active enjoyment (or hedonism)
7. Happiness is the quiet mind, wise and righteous living

A modern stoic would correspond to the responsible, duty-minded, hard-working, public-spirited American who feels obligated to do a good job for his government, company, or other institution, and works well without necessarily getting much recognition, supervision, or special reward.

One might distinguish between "Greek" epicureanism, which would correspond to the above, and "modern" epicureanism, which would not emphasize lines 6 and 7 as much. So far as modern epicureanism is concerned, we might wish to distinguish among "wholesome" or "square" epicureanism (e.g., the sophisticated portion of the Southern California "Bar-B-Q" society), the "hip" epicureanism of the "consciousness-expanding," "dropout," "turned-on," and so-called "joy-love" cultures, and hedonistic or aesthetic epicureanism—using these last terms in their current usage.

Finally we can think of the "gentleman" and the "humanist" as being basically interested in self development—the former in acquiring socially approved skills and experiences and having a sense of noblesse oblige to the state, the latter being more interested in idiosyncratic skills and experiences and being committed to more universal values.

The following tables represent one necessarily quite impressionistic (and perhaps somewhat idiosyncratic) attempt to contrast some characteristics that the Americans have in common with the ancient Romans and the Europeans with the ancient Greeks, especially the Athenians and other Ionic Greeks. (The Spartans, of course, were much closer to the Romans.) In some ways the Britons of the eighteenth and nineteenth centuries were even more like the Romans than the Americans are (e.g., the aristocratic element, the colonial conquests, the even more stringent supremacy of Law), and the Prussians of the eighteenth century very closely resemble the Spartans; that is, they were much closer to the Romans than to the Athenians.

The first table tries to point out that the analogy is incomplete—even at the "loose generalization" level—by presenting obvious dissimilarities between the democratic Americans and the aristocratic Romans. The second table lists some aspects of the basic analogy that we will occasionally use to make or illustrate some point.

TABLE IV.   Some differences between Americans & Romans

Characteristic	American Middle Class	Roman Middle/Upper
Human equality	Axiomatic	Absurd
World view	Optimistic	Pessimistic (Stoic or Epicurean)
Dignity	"Putting on airs"	Basic virtue
Poverty	Vice	Basic virtue
National memory	Extremely short	Extremely long
World domination	Still unwilling to accept	Eventually eager to accept

TABLE V.   Typical attitudes of elites

Characteristic	Romans & Americans	Ancient Greeks & Modern Europeans
Power evokes	Respect	Suspicion
Highest political force	Rule of law	Personality
Most great men receive	Admiration	Denigration
Individualism	Subordinated	Untamed
The primitive masses are	"Citizens"	"Barbarians" or "Helots"
Educated aliens evoke	Slight awe	Amused surprise
Hero Image	Puritan, virtue, stern, conservative, devoted to duty (Cato Sr., George Washington)	Adventurous, unpredictable, conqueror of men and women (Odysseus, Napoleon)
Areas of greatest proficiency	Technical; plumbing, civil & military	Theoretical; philosophy, mathematics, pure science, the arts
Greatest weakness	Theoretical knowledge as opposed to applied	Unity, collaboration
Dillettantism	Unappreciated	"Gentlemanly"
Attitude to the other's culture	Indebtedness	Disdain
Leisure	A vice	A goal
The ambition of others	Highly approved, unless it becomes a threat to the Republic	A threat to one's own and therefore disapproved

Two points are worth noting about such generalizations. While they are often abused, nevertheless they have some value, while they have some value, they are by no means broadly or strictly applicable. For example, there are many notable exceptions among Romans, Americans, Greeks, and Europeans. The Spartans, the late British ruling class, the Prussians, and the Communists actually partake much more of the characteristics of the Romans in the above list than they do of those listed as typically Greek. This applies also to many members of French elites. There are also many other obvious respects in which the Americans differ from the Romans and in which modern Europeans differ basically from the ancient Greeks. The differences between groups, and within each group, are obvious and substantial. What is less—to the modern reader at least—obvious, and yet perhaps of some interest, is that useful parallels can be discerned at all, and will be acknowledged by many knowledgeable people—though uncomfortably, since we have all learned, quite properly, to distrust "generalizations." The great interest of the analogy is that in many ways Europe and Greece and America and Rome, started from similar points and, most important, something very much like our multifold trend occurred in Hellenistic Greece, the late Roman Republic, and the early Roman Empire.

D. ALIENATION AMIDST AFFLUENCE

In this chapter we are attempting to describe a plausible and culturally consistent projection of a culture, values, and style of life consistent with other features of our Standard World of the year 2000. To test whether such a projection is plausible and consistent is both naïvely simple and insolubly complicated. We can take what we now know about past and current American styles of life together with some current trends—and our knowledge of these is far from complete—and add what we believe or find plausible about the socialization of the child, the development of character, character changes in later life, ways in which social structure and culture change, and so forth, and on this basis try to assess the consequences of some simple, basic trends that are characteristic of our Standard World. These include relatively easy affluence, new technology, absence of absorbing international challenges, and considerable but not disastrous population growth. We must ask, in effect, how these trends might furnish or constrain possibilities for change in the large number of Americans already living who will probably survive into the year 2000, and in those who will be born and "socialized" in the interim.

The first salient factor seems likely to be a vastly increased availability of goods and such services as transportation and communication. A second is a likely increase in leisure and a concomitant reduction of the pressures of work. A third is the likelihood of important technological changes in such areas as psychopharmacology, with possible radical consequences for culture and styles of life. Perhaps the most important is a likely absence of stark "life and death" economic and national security issues.

How can we assess the impact of these changes even on a current situation which itself is imperfectly understood? One of the greatest problems of all psychological and sociological speculation has to do with the dialectical quality of the processes involved. It is difficult to know whether to extrapolate trends or to postulate reactions against the same trends. For example, if work will occupy fewer hours of the average person's life, it is plausible to speculate that for this reason work will become less important. On the other hand, it is at least equally plausible that the change in the role of work may cause work as an issue to come to new prominence. The values surrounding work, which in the developed areas have evolved over centuries, may emerge into a new flux and once again become controversial sources of problems within a society and for many individuals. The ideologies that surround work and give it justification and value, in individual and social terms, may become strengthened in support of what remains of work; on the other hand, they may increasingly come into doubt and become the objects of reaction and rebellion. Indeed both trends may materialize simultaneously in different parts of society and may cause conflicts within many individuals. Clearly one can write many scenarios here, with many different branching points. These quandaries must be resolved ultimately by at least partly intuitive and subjective judgments; the most one can claim for such speculations is that no alternative possibilty seems much more likely.

1. Economic Plausibility and Postindustrial Leisure

Let us assume, then, with expanded gross national product, greatly increased per capita income, the work week drastically reduced, retirement earlier (but active life-span longer), and vacations longer, that leisure time and recreation and the values surrounding these acquire a new emphasis. Some substantial percentage of the population is not working at all. There has been a great movement toward the welfare state, especially in the areas of medical care, housing, and subsidies for what previously would have been thought of as poor sectors of the population. Tables VI and VII show one possibility or "year 2000 scenario" for the distribution of work and leisure.

TABLE VI.   A leisure-oriented "postindustrial" society
(1100 working hours per year)

7.5 Hour Working Day
4   Working Days per Week
39   Working Weeks per Year
10   Legal Holidays
3   Day Weekends
13   Weeks per Year Vacation
(Or 147 Working Days and 218 Days Off/Year)

TABLE VII.   Thus in a leisure-oriented society one could spend:

40 per cent of his days on a vocation
40 per cent of his days on an avocation
20 per cent (or more than 1 day per week) on neither—that is, just relaxing

In Chapter III we projected economic indices for the United States in some detail, and indicated how extrapolations to very prosperous levels of personal income were consistent with reasonable assumptions about rates of increase in labor productivity and substantial reduction in hours worked and labor force participation rates. One might construct additional "quantitative scenarios." For example, a projection for a leisure-oriented society, without alienation, could retain the population of 318 million but cut the work year from 1600 to 1100 hours (1920 hours is new "standard" but somewhat above average) and lower the labor force participation rate to 56 per cent. Under these assumptions, and assuming that the economy has been experiencing the high rate of productivity increase, the leisure-oriented society will still provide for an increase of about 100 per cent in per capita GNP relative to 1965. Table VIII shows this.

TABLE VIII.   Year 2000 economic scenarios for U.S. affluence and alienation

Population	318 million
Employed Labor Force	122 million
Leisure Oriented Society:	Work year 1100 hours
GNP	$2,321 billion
Per Capita GNP	$7,300

Let us consider in more detail what might happen to the 40 per cent of persons who are normally in the labor force. One possibility for participation rates is set forth in the following table:

TABLE IX.   In a "normal" postindustrial, affluent society of those (40 per cent) normally in the labor force:

50%	Work normal year
20%	Moonlight
10%	"Half-time hobbyists"
5%	Frictional unemployment
5%	Semifrictional unemployment
5%	Revolutionary or passive "dropout"
5%	"Voluntarily" unemployed
100%	

The above is not a very serious conjecture, but it seems not implausible that one-half the people would work in more or less normal fashion, and that one-fifth of the people would work longer hours than normal, either for income or for compulsive or altruistic reasons. Because of the excess contribution of this group, it may be possible to maintain something close to the high GNP projected above, even though some 20 to 30 per cent of the normal labor force contribute little or no labor. The underproducers might be, in effect, hobbyists working a few days a month, or a few months a year, to acquire the income to pursue their hobbies. One can also assume that "normal" frictional unemployment will be somewhat higher than usual, and that there will also be something which might be considered "semi-frictional" unemployment (that is, people who have lost jobs but are taking some time looking for another by using their vacations, or who have unusually high or unrealistic standards of what their jobs should be or who are just lying around living on savings.) There could also be a group, assuming the above conditions, who reject any sort of gainful employment on the ground of principle or preference. And finally, there should be people who are more willing to be on relief than not, if only because they have personal or family problems that make it unwise for them to work if they can survive without; or there may be some who are simply and cynically "on the dole."

The above suggests that in place of the current 20 per cent poor, we may have a similar number, but differently situated, who do not participate normally in the vocational life of the nation.

Consider now the problem of the annual number of hours of work. There could easily be either a four- or five-day week as described in the following two tables:

TABLE X. Some assumed five-day week working patterns

Nominal Hours Per Week	Legal Holidays	Weeks Off	Total Work Days	Total Days Off	Total Hours
5 x 8 = 40	10	2	240	124	1920
5 x 8 = 40	10	4	230	134	1840
5 x 8 = 40	10	6	220	144	1760
5 x 7.5 = 37.5	10	5	225	139	1687
5 x 7 = 35	10	4	230	134	1610
5 x 7 = 35	10	6	220	144	1540
5 x 7 = 35	10	8	210	154	1470

It is difficult to guess what patterns will be assumed. It should be noted that if any of the four-day week patterns were adopted the "normal" worker could spend less than 50 per cent of his days on his vocation (but only seven to seven and one-half per day), less than 50 per cent of his days on an avocation (possibly working somewhat longer than seven and one-half hours per day), and then still

TABLE XI.   Some assumed four-day week working patterns

Nominal Hours Per Week	Legal Holidays	Weeks Off	Total Work Days	Total Days Off	Total Hours
4 x 7.5 = 30	10	4	184	180	1380
4 x 7.5 = 30	10	6	176	188	1320
4 x 7.5 = 30	10	8	168	196	1260
4 x 7.5 = 30	10	10	160	204	1200
4 x 7.5 = 30	10	12	152	212	1140
4 x 7.5 = 30	10	13	144	220	1080
4 x 7 = 28	10	4	184	180	1208
4 x 7 = 28	10	8	168	196	1096
4 x 7 = 28	10	13	144	220	984

have one or two days off a week for just relaxing. In other words, it would be possible to pursue an avocation as intensely as a vocation and still have a good deal of time for "third-order" pursuits. As we pointed out at the beginning of Chapter III, such patterns can be consistent with continued economic growth at reasonably high rates.

2. Success Breeds Failure: Affluence and the Collapse of Bourgeois Values

John Maynard Keynes addressed himself to this dilemma in one of the earliest and still one of the best short discussions of some of the issues raised by the accumulation of wealth through investment. As he put it,

> . . .the economic problem, the stuggle for subsistence, always has been hitherto the primary, most pressing problem of the human race. If the economic problem is solved, mankind will be deprived of its traditional purpose.
>
> Will this be of a benefit? If one believes at all in the real values of life, the prospect at least opens up the possiblity of benefit. Yet I think with dread of the readjustment of the habits and instincts of the ordinary man, bred into him for countless generations, which he may be asked to discard within a few decades. . .thus for the first time since his creation man will be faced with his real, his permanent problem—how to use his freedom from pressing economic cares, how to occupy his leisure, which science and compound interest will have won for him, to live wisely and agreeably and well.

There are those who would argue that with increased freedom from necessity men will be freed for more generous, public-spirited, and humane enterprises. It is a commonplace of the American consensus that it is poverty and ignorance that breed such evils as Communism, revolutions, bigotry, and race hatred. Yet we know better than to expect that the absence of poverty and ignorance will result in a triumph of virtue or even of the benign. On the contrary, it is equally plausible that a decrease in the constraints formerly imposed by harsher aspects of reality will result in large numbers of "spoiled children." At the minimum many may become uninterested in the administration and politics of a society that hands out "goodies" with unfailing and seemingly effortless regularity.

One may choose almost at will from among available hypotheses that may seem to apply to the situation, and one reaches contrary conclusions depending upon the choice that is made; this indeterminancy is perhaps a measure of the inadequacy of contemporary social thought as a basis for generalization, relative to the complexity of human phenomena.

For example, one may take the Dollard *et al.* frustration-aggression hypothesis and conclude that aggressiveness will be greatly tranquilized in a society that provides much less external and realistic frustration. This is opposed to the more complex and more psychoanalytically oriented point of view of Freud who points to the role that frustrations imposed by external reality may play in shoring up the defenses of the character structure—defenses that are crucial strengths and that were acquired through learning, with difficulty, as an infant to defer gratification and to mediate among conflicting energies of instinctual impulses, conscience, and the opportunites and dangers of the real world.* Research might show, if research could be done on such a subject, that many an infantile and narcissistic personality has matured only when faced with the necessity of earning a living—others only when faced with the necessity for facing up to some personal challenge, such as military service or participation in family responsibility. (The well-known finding that suicide rates drop sharply during wars and economic depressions is subject of diverse interpretation, but it may suggest that such external challenges can serve crucial integrative or compensatory functions for some personalities, and perhaps, less dramatically, for many others.) This is not to say that equally effective or perhaps superior external challenges could not be found to substitute for the working role—or wartime experience—as a maturing or reality-focusing influence. If they are not found, however, while the economy and international and other threats make fewer demands, the decline of the values of work and national service may have some destructive effect.

Thus there may be a great increase in selfishness, a great decline of interest in government and society as a whole, and a rise in the more childish forms of

---

*As Freud pointed out, "Laying stress upon importance of work has a greater effect than any other technique of living in the direction of binding the individual more closely to reality; in his work he is at least securely attached to a part of reality, the human community. . .and yet. . .the great majority work only when forced by necessity, and this natural human aversion to work gives rise to the most difficult social problems." *Civilization and Its Discontents* (London: Hogarth Press, 1930), p. 34, note I.

individualism and in the more antisocial forms of concern for self and perhaps immediate family. Thus, paradoxically, the technological, highly productive society, by demanding less of the individual, may decrease his economic frustrations but increase his aggressions against the society. Certainly here would be fertile soil for what has come to be known as alienation.

The word alienation has been used in many different senses, some of them well defined and some in the context of systems of explanation and prescription for the ailment. The young Karl Marx, for example, followed Ludwig Feuerbach (and to some extent anticipated Freud's *Civilization and its Discontents*) in the belief that alienation resulted from civilized man's "unnatural" repression of his instinctual, especially sexual, nature. Later, however, Marx concluded that alienation resulted from the worker's relationship to labor that had to be done for the profit of another; the cure was to have the worker "own" the means of production; thus alienation could be reduced by shortening the working day, and "the worker therefore feels himself at home only during his leisure."

The alienation that we speculate may result from affluence could have little or nothing to do with whether the society is capitalist or socialist. In either case the control of the decision-making apparatus would be perceived as beyond the reach of and in fact of little interest for the average person. Thus, whatever the economic system, the politics (and even the culture) of plenty could become one not of contentment but of cynicism, emotional distance, and hostility. More and more the good life would be defined in Epicurean or materialistic, rather than Stoic, or bourgeois terms. The enhancement of private values combined with the increased sense of futility about public values would also entail a kind of despair about the long-run future of the whole society. More and more people would act on the aphorism currently attributed to a leader of the new student left: "If you've booked passage on the Titanic, there's no reason to travel steerage."

Thus the classical American middle-class, work-oriented, advancement-oriented, achievement-oriented attitudes might be rejected for the following reasons:

1. Given an income per worker by the year 2000 of well over ten thousand dollars in today's dollars, it may become comparatively easy for intelligent Americans to earn ten to twenty thousand dollars a year without investing very intense energies in their jobs—in effect they will be able to "coast" at that level.

2. It may become comparatively easy for an American to obtain several thousand dollars a year from friends and relatives or other sources, and to subsist without undergoing any real hardship, other than deprivation of luxuries. (Informal polls in the Cambridge, East Village, and Haight Ashbury areas indicate that many "hippies" get along on about ten dollars per week, as do many CORE and SNCC workers.)

3. Welfare services and public facilities will generally probably put a fairly high "floor" under living standards, even in terms of luxuries such as parks, beaches, museums, and so on.

4. With money plentiful, its subjective "marginal utility" would probably tend to diminish, and there would probably be a greatly increased emphasis on things that "money cannot buy."

5. Economic and social pressures to conform may diminish as the affluent society feels increasingly that it can "afford" many kinds of slackness and deviation from the virtues that were needed in earlier times to build an industrial society.

6. If the "Puritan ethic" becomes superfluous for the functioning of the economy, the conscience-dominated character type associated with it would also tend to disappear. Parents would no longer be strongly motivated to inculcate traits such as diligence, punctuality, willingness to postpone or forego satisfaction, and similar virtues no longer relevant to the socio-economic realities in which children are growing up.

7. Yet the need to "justify" the new patterns may remain, and to the extent that there is any residual guilt about the abandonment of the nineteenth- and early twentieth-century values, there would be exaggerated feelings *against* vocational success and achievement. Many intellectuals and contributors to popular culture would help to make the case against "bourgeois," "managerial," "bureaucratic," "industrial," "Puritanical," and "preaffluent" values. There would then be considerable cultural support for feelings ranging from indifference to outright contempt for any sort of success or achievement that has economic relevance.

Other factors would augment these effects. For example, presumably by the year 2000 much more will be known about mood-affecting drugs, and these drugs will probably be used by many as a means of escape from daily life. At the same time, the young, those without responsibility in the social system, will be increasingly alienated by a society that conspicuously fails to meet what it judges to be minimal standards of social justice and purpose (standards which look impossibly Utopian to decision-makers). Ideological movements would form to rationalize and justify rebellion and renunciation of old "obsolete" values by youth from all classes and strata of society. Less articulate but equally rebellious young people would contribute to a great rise in crime and delinquency. Other symptoms of social pathology, such as mental illness, neurosis, divorce, suicide, and the like would also probably increase. Traditional religious doctrines might either continue to lose force or continue to be reinterpreted, revised, and secularized so as to pose few obstacles to the current general way of life.

On the other hand, the resources of society for dealing with these problems, perhaps in a (suffocatingly?) paternalistic way, would also have been greatly augmented. Before discussing the differences that might be made by social responses to these problems, let us see how they might affect various social groups.

E. ALIENATION AND THE SOCIAL STRUCTURE

Of course not everyone would suffer equally from the prevalence of affluence that we have just described. Among the voluntary poor, for example, there would be certain rock-bottom types who would insist on deprivation for reasons that have to do with personal psychopathology. Thus many skid-row derelicts, alcoholics, drug addicts, ambulatory schizophrenics, and other marginal or self-destructive personalities would insist on living at a level barely sufficient for survivial. Some, indeed, would insist on slow forms of suicide through starvation, exposure, or malnutrition, as with some cases of alcoholism.

Most of the relatively poor members of society would, however, be amply subsidized. They would readily accept welfare as a means of support, and the feeling that the world owes them a living would go largely unquestioned. Incentives to take unskilled jobs would be minimal, nor would holding a job—particularly a marginal one—add much to self-esteem when relief and welfare have so much group approval. Extremist movements might flourish in the general climate of alienation from the "power structure." Many whites and middle-class Negroes might view race riots and acts of destruction with indifference, or even sympathy and approval. The following statement of a well-known poet of Negritude may well come to reflect the sympathies of a larger segment of both populations:

> Mercy! mercy for our omniscient conquerors
> Hurray for those who never invented anything
> Hurray for those who never explored anything
> Hurray for those who never conquered anything
> Hurray for joy
> Hurray for love
> Hurray for the pain of incarnate tears.*

At the same time, since what Oscar Lewis has described as the "culture of poverty"—with its short-time perspectives and emphasis on immediate survival and pleasure, and the like—would have become, to some degree, also the culture of affluence, the assimilation of the impoverished ghetto-dweller into the larger

---

*Aimé Césaire, *Cahier D'un Retour Au Pays Natale,* as quoted in Colin Legum, *Pan-Africanism,* rev. ed. (London: Pall Mall Press Ltd.; New York; Frederick A. Praeger, 1965).

society would pose less difficult psychological problems. The indolent spectator, the "hipster" and the "swinging cat" would have become in large degree the norm for very wide sectors of the population. Moreover, this group would be receptive to ideologies which welcome the downfall and dissolution of the American postindustrial way of life. These people would tend to be congregated in the major cities that they would probably not control politically but in which they would constitute major pressure groups and could exercise veto rights on many programs. They would probably live in rather uneasy and unstable alliance with the upper middle class, "responsible" people who would continue to control the economic structure and make use of the resources of the city.

The following tables indicate one estimate (from *U.S. News and World Report*) of the likely increase in the concentration of Negroes in major cities. These figures seem to be those that might be reached by a straightforward projection of current trends and probably exaggerate the proportions of the

TABLE XII(a).   How Negro population will grow in the cities

Negroes will make up half or more of the population inside these cities by the year 2000:

	Per Cent Negro in 1960	Per Cent Negro in 2000
Washington, D. C.	53.9	75
Cleveland, Ohio	28.6	67
Newark, N. J.	34.1	63
Baltimore, Md.	34.7	56
Chicago, Ill.	22.9	55
New York City	14.0	50
Philadelphia, Pa.	26.4	50
Detroit, Mich.	28.9	50
St. Louis, Mo.	28.6	50

Negroes will make up one-third to one-half of the population inside these cities:

	Per Cent Negro in 1960	Per Cent Negro in 2000
Atlanta, Ga.	38.3	44
Kansas City, Mo.	17.5	42
Cincinnati, Ohio	21.6	40
San Francisco-Oakland, Calif. . .	14.3	40
Houston, Tex.	22.9	34
Buffalo, N. Y.	13.3	34
Pittsburgh, Pa.	16.7	34
Paterson-Clifton-Passaic, N. J. . .	9.3	34

Negroes will make up one-fifth to one-third of the population inside these cities:

	Per Cent Negro in 1960	Per Cent Negro in 2000
Boston, Mass.	9.1	31
Dallas, Tex.	19.0	30
Milwaukee, Wisc.	8.4	29
Los Angeles, Calif.	12.2	20

TABLE XII(b).   How Negro population will grow in the suburbs

In suburbs surrounding big cities, Negro population will grow in years ahead—but still will remain less than one-fourth of the total. Here, for some major cities, are the percentages of Negroes in the suburban population:

	Per Cent Negro in 1960	Per Cent Negro in 2000
San Francisco-Oakland, Calif. . .	4.8	22
Washington, D. C.	6.1	19
Baltimore, Md.	6.7	19
Philadelphia, Pa.	6.1	18
Detroit, Mich.	3.7	13
Houston, Tex.	10.3	12
Cincinnati, Ohio	3.4	12
Cleveland, Ohio	.7	12
Newark, N. J.	6.7	10
Chicago, Ill.	2.9	10
San Diego, Calif.	1.1	10
St. Louis, Mo.	6.1	7
Paterson-Clifton-Passaic, N. J. . .	1.9	7
Kansas City, Mo.	5.9	6
Atlanta, Ga.	8.5	6
Dallas, Tex.	6.5	5
Pittsburgh, Pa.	3.4	4
Los Angeles, Calif.	3.1	4
Boston, Mass.	.8	2

ethnic shift that is in prospect. (Unless these projections assume *all* Negroes have moved to the listed areas, they imply an implausibly high increase in the percentage of Negroes in the total United States population by the year 2000.) Even so, they are an indication of already obvious changes in urban demography that may have great political and economic significance.

TABLE XII(c).   How Negro population will grow in total
metropolitan areas

Including both suburbs and central cities, Negroes by 2000 will make up one-fifth or more of the population in these metropolitan areas:

	Per Cent Negro in 1960	Per Cent Negro in 2000
Baltimore, Md.	21.9	30
Washington, D. C.	24.3	27
Philadelphia, Pa.	15.5	27
New York City	11.5	27
Cleveland, Ohio	14.3	26
Chicago, Ill.	14.3	25
San Francisco-Oakland, Calif. . .	8.6	24
Detroit, Mich.	14.9	20
Atlanta, Ga.	22.8	20
Houston, Tex.	19.8	20

Negroes will make up one-tenth to one-fifth of the population in these metropolitan areas:

	Per Cent Negro in 1960	Per Cent Negro in 2000
Cincinnati, Ohio	12.0	19
Newark, N. J.	13.3	18
St. Louis, Mo.	14.3	17
Kansas City, Mo.	11.2	16
Dallas, Tex.	14.3	13
San Diego, Calif.	3.8	13
Buffalo, N. Y.	6.3	11
Paterson-Clifton-Passaic, N. J. . .	3.6	10

SOURCE: Census Bureau, 1960; estimates by USN&WR Economic Unit for year 2000, copyright 1966, *U.S. News & World Report, Inc.*, from issue of February 21, 1966.

It has often been pointed out that current United States racial conflicts involving issues such as housing and education are based as much on issues of poverty and social class as on racism. Although the racism and militancy of the Black Muslims (and the possibility that they may be misused as a group by those who eventually succeed in what seems to be an intermittent power struggle within the movement) may make it difficult for whites to welcome the group, it may be the most active force since the Christian evangelists in recruiting new Negro members into the middle class.

TABLE XIII.   The Black Muslims

1. Become legally married
2. Give up alcohol, drugs, and pork
3. Work 8 hours per day—6 days per week
4. Support their families
5. Are strong (even domineering) father figures
6. Dress neatly and conventionally
7. Send their children to school
8. Maintain their dignity and self-respect
9. Save their money
10. Often become small entrepreneurs
11. Hate whites and preach apartheid and violence

Thus, perhaps suprisingly, this group may be the means—over one or two generations—for many who might otherwise have remained in the "culture of poverty" to acquire "father figures," motivation for striving, and ultimately typical American middle-class attitudes and occupations. When this process has been completed, the racism of the Nation of Islam may disappear or diminish, leaving little more "unmelted" in the "pot" than in the typical "successful" American pattern, in which there ordinarily remains some ethnic self-consciousness (and some "intergroup tensions") among mostly assimilated, but still distinct, former immigrant groups.

Returning to the society as a whole: The lower middle classes (who in general will be making between ten and twenty thousand 1965 dollars a year) would enjoy a greatly reduced work week with some emphasis on leisure. While their necessities and basic luxuries would be obtainable without great effort, they might still wish to increase income by moonlighting or by the wife's working. Some, of course, would have little motivation for expending extra effort and for them the problems of occupying leisure time would be a primary concern. Others would want to save money, pursue expensive hobbies, or emulate some aspect of the life patterns of the upper middle classes or even the wealthy. Both groups would provide a tremendous market for all kinds of sports and fads and particularly for various forms of mass entertainment. Year 2000 equivalents for the bowling alley, miniature slot-racing car tracks, and the outboard motor, would be everywhere. The drive-in church, the "museum-o-rama," and comparable manifestations of pressures toward a general degradation and vulgarization of culture would be a likely result of the purchasing decisions of this group. At the same time, these people might militate politically against civil rights and against the poor and relatively poor nonworking classes that they must support, and they would likely provide the primary support for both conservative national policies and political jingoism.

The upper middle class (most of whom will have annual income of perhaps twenty to sixty thousand 1965 dollars), by contrast, would, in many ways, be emulating the life-style of the landed gentry of the previous century, such as emphasizing education, travel, cultural values, expensive residences, lavish entertainments, and a mannered and cultivated style of life. For some there would be much effort to amass property and money for personal and family use. Getting away from the cities and from centers of population would be a difficult problem which only large amounts of money will solve. There would probably be some emphasis on "self-improvement" including cultural dilettantism. While among most members of this group we would expect a continuation of current well-to-do suburban patterns, in many cases patterns of life might be increasingly self-indulgent, marriages unstable, children alienated from their parents. Interest in strange and exotic political ideologies, Eastern mysticism, and the like, might flourish, as could a cult of aestheticism and a shrinking from the "grubby" or "crass" aspects of society. Effete attitudes might be combined with contempt for the lower middle class and fear of the poor and of their propensity for violence. There may also be some romanticization of the "noble savage" (or "hippy") who lives outside the values of the society, in voluntary poverty and/or minor or even major criminality.

The very wealthy would be able to buy considerable protection from these exigencies—that is, from all but the cultural confusion and normative conflicts. Because of their social power, many would have responsibility and there might be, in some groups, a sense of *noblesse oblige,* which would be shared by many in the upper middle class.

Youth could be especially self-indulgent or alienated, as the identity confusion typical of adolescence is exacerbated by the confusion, normlessness, and anomie of the society. Indifference to moral and ethical values and irresponsibility of personal behavior would be combined with feelings of outrage about the vast discrepancies between the wealth of the rich nations and the poor, and an especially painful situation would arise if these young people were drafted for military service in teeming, underdeveloped countries. Combined with pacifism and antipatriotic ideologies would be a strong feeling that American lives are too precious to be spent anywhere else in the world—or indeed to be wasted in America itself. Recruitment into any of the more difficult or demanding professions would be restricted to those (perhaps many) who have adopted Stoic patterns, and to the sons of fathers who are already in those professions and who identify with them. Conformers would—as always—work, aspire to comfortable sinecures, and look forward to early retirement—but now with great confidence. "Bumming around" and hip patterns of life could become increasingly common (though not the norm) in all but the lower middle-class groups. Many would live indefinitely on the resources of friends and relatives and on opportunistic sources of income without doing any sustained work, or in the upper middle-class pattern would cloak themselves in pretentions to artistic creativity. In spite of the prominence of symbols of rebellion and nonconformity, these youths, especially because of their anomie and alienation, would be subject to extreme fads of behavior and political, ethical, and religious ideas.

Of course, it is important to note that the lower middle class, making, say, five to twenty thousand dollars a year, are by 2000 not going to be very different from the lower middle and middle middle classes today. The upper (and middle) middle class in the year 2000 make, say, twenty to one houndred thousand dollars a year and will not necessarily feel independently wealthy. Both groups by and large will probably continue (with some erosion) with current work-oriented, advancement-oriented, achievement-oriented values. The extreme alienation we are talking about is restricted to minorities, which will be important in part because they are likely to be concentrated in the big cities, in part because they appeal to many of the more intellectual members of that ordinarily alienated group—adolescents—and mostly because their members will be literate and articulate and have a large impact on intellectuals and therefore on the culture generally.

It would be useful to make explicit the notions that determine these speculations and to discuss the alternative speculations that might be made. In particular, the evidence that might be found for or against the various alternative hypotheses and speculations should be given; only a small start can be made in this direction here.

F. FUNCTIONS OF WORK

To arrive more precisely at an answer to the question, "What are the consequences of a reduction in the amount of work that needs to be done?" one must ask, "What are the various functions for the individual of the work he performs?" It is easy to make a long list of such benefits at various levels of analysis. For example, people derive from work such benefits as role; status; sense of striving; feeling of productivity, competency, and achievement; and relationships with others and advancement in a hierarchy, whether organizational or professional.

The following table shows some rough characterizations and generalizations about various roles work may play for different kinds of people in the year 2000. Those whose basic attitude toward work is that it is totally abhorrent or reprehensible are not listed, since on the whole they will find it possible to avoid employment entirely.

As discussed later one could easily imagine that many Americans from "normal" (i.e., not deprived) backgrounds will increasingly adopt the first

TABLE XIV.

Basic Attitude Toward Work As:	Basic Additional Value Fulfilled by Work
1. Interruption	Short-run income
2. Job	Long-term income—some work-oriented values (one works to live)
3. Occupation	Exercise and mastery of gratifying skills—some satisfaction of achievement-oriented values
4. Career	Participating in an important activity or program. Much satisfaction of work-oriented, achievement-oriented, advancement-oriented values
5. Vocation (calling)	Self-identification and self-fulfillment
6. Mission	Near fanatic or single-minded focus on achievement or advancement (one lives to work)

position, that work is an interruption, while many formerly in the lower and economically depressed classes will increasingly shift to the second or third positions which reflect more work-oriented and achievement-oriented values. On the other hand, the man whose missionary zeal for work takes priority over all other values will be looked on as an unfortunate, perhaps even a harmful and destructive neurotic. Even those who find in work a "vocation" are likely to be thought of as selfish, excessively narrow, or compulsive.

Many of the benefits of work could be derived from other forms of activity, provided they were available and, preferably, institutionalized. The model of the cultivated gentlemen, for example, is likely to be available and possibly generally usable in a democratic and upward mobile society like the United States. It may be argued that aristocrats are far more visible in Europe and that it is more respectable for the wealthy to live as landed gentry, rentiers, even as playboys in Europe than in the United States, and that for these reasons the transition to this pattern of life would be easier in Europe. Indeed, historically this has often been the aspiration and achievement (after a generation or two) of the upper middle class and even the lower class nouveaux riches. On the other hand, if it became the ideal it is probably more difficult for a typical European to think of making the social transition to such a status for himself than it would be for the typical American. Of course, the American has seen fewer examples of such lives, and has up to now respected them less. Therefore it seems less likely to be the American ideal.

In the economic structure we are describing, there may be a special problem of the service professions whose productivity per hour may not have gone up. Thus many believe there are probably important limits to the extent to which the efficiency of persons such as teachers, professors, doctors, lawyers, ministers, psychologists, social workers, and so forth can be increased. Others believe that not only can these professions be automated,* but that there are huge opportunities for increasing efficiency through better organization, specialization, and the very skilled use of computers. Nevertheless, there are likely to remain irreducible kinds of activities that defy rationalization or improvement, such as those that require face-to-face meeting and conversation. Thus programmed instruction, lectures, and sermons over television are not likely to displace face-to-face human communication, at least not without great loss to those involved. Therefore only part of the current activity in these fields is likely to increase in productivity.

To the extent that recruitment into the service professions is greatly expanded because of the reduced need for people in manufacturing, routine aspects of public administration, and automated administrative and managerial tasks, several problems will arise. One is that it will be perhaps more difficult to recruit people to do difficult and demanding work that either requires long and arduous training or requires working under difficult, dangerous, or frustrating conditions. If the hours of work of people in these professions go down severely, the incentives and psychological functions of membership in the profession may be somewhat diluted. For example, a hospital may have three head nurses if there are three shifts; what happens, however, when there are six or eight shifts? To what extent is authority, expertise, and satisfaction diluted when power, responsibility, and status are so fractionated?

Similar questions should be posed about other kinds of activities. In general, a threefold increase in GNP per capita is far from the equivalent of a threefold increase in productivity per capita in all relevant respects. As real productivity increases dramatically in certain industries, principally in manufacturing and heavily clerical industries, such as banking and insurance and many federal, state, and local governmental functions, which could be very much automated, the price structure would also change dramatically. This would result in enormous increase in the availability, variety, and quality of goods and many standardized services, since these items would become very much cheaper or very much better for the same price. A threefold increase in GNP per capita would probably imply a much greater increase in standard of living with respect to these items. Yet, at the same time, skilled, personal services requiring irreducible quantities of human time, training, and talent would become both absolutely and relatively

---

*Thus much legal research can be done most easily through a computerized library. A physician may be able to phone a list of symptoms into a central computer and get back a print-out of suggested diagnostic possibilities. Many laboratory tests might be performable by methods which would present immediate results. Closed circuit television and various kinds of continuously reading tests presented on central display boards could even now make the utilization of hospital personnel also much more efficient. Other possibilities were mentioned in Chapter II.

expensive. Thus there would probably still be a very strong demand for, and probably also a much expanded supply of expensive and skilled professionals, managers, entrepreneurs, artisans, technicians and artists—for the most part, the well-educated upper middle class. This group may well be much too busy and well rewarded to be alienated.

Furthermore, even if one imagines the ordinary member of the labor force amply supplied with intricate technology affording innumerable needs and luxuries during his short work week, and even if he can travel anywhere in the less-developed world and easily buy vast quantities of domestic service and other personal attentions during his long vacations, many important consumer items are likely to remain too expensive for him to wish his work week to become *too* short. There will probably still be a class of "luxury items," consisting of such things as vacation houses in extremely exotic places, advanced or "sporty" personal vehicles such as perhaps ground-effect machines, or similar items for the most part well beyond today's technology and prohibitively expensive for ordinary workers by today's standards, that by the year 2000 will be still expensive, but perhaps within reach of the man who is interested in earning enough money—and many, no doubt, will be interested.

G. OTHER FACTORS IN ALIENATION

In discussing alienation, attention ought also to be given to other aspects of cultural change that may contribute to ego-disintegration and feelings of disorientation. Here we meet the difficult problem of diagnosing the malaise of our times. What precisely causes the alienation of adolescents and many others in 1967 is a very controversial matter. Speculation about the year 2000 is, of course, many times more complicated and uncertain. If it sometimes seems to be easier this is presumably mostly because the results are less checkable—though it is possible that there will be clear and predictable tendencies in 2000 that are only ambiguously detectable today. In any case it seems plausible that the "end of ideology" and an inevitable disenchantment with the ideals and expectations of American democracy and free enterprise, coupled with a continued decline in the influence of traditional religion and the absence of any acceptable mass ideologies, have and will continue to contribute to a common spiritual and political rootlessness. As secularization, rationalization, and innovation continue to change the culture in the direction of Sensate and bourgeois norms, the influence of traditional *Weltanschauungen* seems more likely to continue to wane that to undergo any resurgence in the next thirty-three years.

Furthermore some things have happened or are happening that change man's relation to his universe in ways that may be unsettling for many people. For example, the inventions of nuclear weapons and intercontinental delivery systems have probably made human life permanently more precarious, and have introduced into international relations a new level of potential horror that is difficult even to imagine with any precision. At the minimum they provide any who wish for it a good excuse for aimless drifting or horrified resentment; in addition, they are ample reasons for both realistic concern and widespread neurotic anxiety and despair.

Technological change itself may contribute to feelings of estrangement from the new physical world and also from a society strongly affected by continual innovation and disruption. There is a long tradition in American letters of hostility to the machine,* and, at least since World War II, an increasing perception that the social consequences of science and technology are, at best, mixed blessings. Machines that perform some functions of the human mind far better than humans can are likely to be even more resented, in spite of their economic benefits, than machines that do the same for human muscles. The human place in the world may be most seriously disturbed by new medical technology. New drugs will raise sharply the questions, what is a real human feeling, and what is a genuine personality? Plastic replacements for hearts and other vital organs raise in new and more difficult form the old problem of defining life and death, and add a new difficulty to the old question, what is a human being?

The exploration of space already under way may also have a somewhat disturbing impact on the imaginations of many people. It is well known, for example, that many schizophrenics are preoccupied with fantasies of space exploration (as well as fantasies involving the immense destructive potential of the H-bomb). Phenomena such as weightlessness, dependence for existence on a wholly artificial and technologically sustained environment, and isolation from familiar objects and human kind may have obvious impact on some minds and more subtle impact on a great many. The unconscious is, as Freud has reminded us, an inveterate punster, and it may not be accidental that phrases such as way out, far gone, out of it, and out of this world are currently used to mean strange or bizarre; and that, moreover, phrases such as way out, dropout, flip-out, freak-out, turned on, tuned in, out of my head, and cool are supposed to refer to desirable conditions. Perhaps the most important alienating influence will be a purely negative thing—the absence of the traditional challenge of work, community approval, and national needs.

---

*Leo Marx in *The Machine in the Garden: Technology and the Pastoral Idea in America* (New York: Oxford University Press, 1964), treats this American tradition at some length and with extreme sympathy. Lewis Mumford's *The Myth of the Machine: Technics and Human Development* (New York: Harcourt, Brace and World, 1967), is the latest in his long series of works belonging to this tradition; in particular, it represents a reinterpretation of archaeology and early history intended to show that speech, and not the ability to use tools and technics, was once, and should become again the essential human faculty. See also Eric Hoffer, *The Temper of Our Time* (New York: Harper and Row, 1967), for some caustic comments on this idiosyncrasy of some intellectuals.

H. HUMANISM AND THE VALUE OF TIME

It is possible to suppose that something else might happen. For example, John Adams, our second President, once suggested that: "my sons ought to study mathematics and philosophy, geography, natural history and naval architecture, in order to give their children a *right* to study painting, poetry, music, architecture, statuary, tapestry and porcelain. . ." (emphasis added).

The passage is peculiarly American; almost no (correspondingly upper-class) European would use the word "right." The most he would have said would be that his sons *ought* to emphasize mathematics, philosophy, geography, and so on, in order that their sons *could* emphasize painting, poetry, music, and the like. He would feel that some interest in painting, poetry, and music was proper and unremarkable. On the other hand for most Americans a man who is deeply preoccupied with porcelain, or any of the fine arts, may still be, even in this less Philistine age, a bit suspect—whether as effeminate or as simply not sufficiently serious and practical. Adams's statement is characteristically American, in that it gives an overwhelming priority to the needs of national security and statesmanship and asserts that no one has a *right* to devote his attention to "finer things" for their own sake, until these needs have been adequately met. A contemporary parallel is the American upper middle-class view of the proper relation between work and play. Typically an American businessman or professional man apologizes for taking a vacation by explaining it is only "in order to recharge his batteries"; he justifies rest or play mostly in terms of returning to do a better job. The European by contrast seems to enjoy his vacation as a pleasure in its own right, and does not hesitate to work for the express purpose of being able to afford to play in better style.

We have already suggested that in the postindustrial society that we are describing, in continental Europe the middle and upper classes could, in effect, return to or adopt the manner of the "gentleman." Many Europeans, of course, argue that things are now going the other way, that under the impact of a mass-consumption, materialistic culture the humanistic values that have been so characteristic in Europe are rapidly eroding or disappearing. One can fully concede that this is indeed the current phase and still note that there is likely to be a reaction in the not-too-distant future. Even today many Europeans seem to emphasize nonvocational aspects of their life much more intensely than even the family-oriented Americans do. For example, many Europeans seem to plan intensely, a year ahead, how they will spend their vacations. Once on vacation, they would resent any interruption for work—such as a business phone call—far more than any American would. It is the vacation that, at the time at least, deserves to be taken more seriously—not the work.

While the American has no sense of "staying in his place" and therefore could seek to emulate aristocratic ideals, the issue remains as to what extent the gentleman of leisure will be an ideal. It simply has not been one in the United States in the last one hundred and fifty years. On the other hand, the middle and upper middle class in Europe have often aspired to be gentlemen, and when tradesmen made fortunes they often made the transition, or at least their children did. In the United States, on the contrary, a member of the middle class who makes his fortune, or his descendants, such as those in the Yankee upper class, usually persists in the tradition of hard work and of service of one kind or another.

Thus if the average American had an opportunity to live on the beach for six months a year doing nothing, he might have severe guilt feelings in addition to a sunburn. If an American wishes to be broiled in the sun, he usually must go through a preliminary justification such as the following: "The system is corrupt, I reject it. Its values are not my values. To hell with these puritanical, obsolete concepts." Only at that point can he relax in the sun. If he is more guilty, or articulate, he may proclaim: "All of those robots who are working have sold out to the soulless, inhumane system with its obsolete and grubby, machine-based, materialistic values, and its empty goals of the Bitch-Goddess of monetary success. By refusing to be drawn in, I at least can perserve my humanity, individuality, integrity, dignity, and physical health, as well as my spontaneity, the freshness of my perceptions, the openness of my relating, and my capacity to love." Unless an American has taken an ideological and moralistic stance against the work-oriented value system, he cannot abandon work.

On the other hand, a good many Americans, and typically middle-class Americans, will have a sense of noblesse oblige. Some of the same pressures toward Stoic values that were important in the Roman Empire will be important here as well.

Let it be added that in this "super-affluent" society of year 2000, it is not likely that efficiency (defined by the criteria of maximizing profit or income) will still be primary, though it will doubtless remain important. To some degree this has already occurred and the situation in the United States is today very different from what it was before 1929. For example, it seems to be true that when a middle-class American looked for a job in 1929, he was interested in salary and prospects for advancement. Today, however, the first questions addressed to personnel interviewers are more likely to relate the satisfaction of the applicant's family with the new neighborhood and the quality of the schools. This is, of course, particularly true of professional and managerial workers, but it seems to be more widely spread as well. It is only after the requirements of home and children have been satisfied (and sometimes considerations of pension, vacation, and insurance as well), that salary and advancement are discussed.

We could think of this phenomenon as a shift to humanistic rather than vocational or advancement-oriented values, and conjecture that this tendency will increase over the next thirty-three years. Indeed, unless there is a surprising interruption in the exponential progress of prosperity, sensate-humanist and epicurean values almost surely will come to dominate older bourgeois virtues, and may even return, in some respects, to criteria that antedated the "bourgeois" element of the multifold trend, which has been a driving force for more than five centuries. Thus Keynes—here the "reactionary"—returns to the Sermon on the Mount:

> I see us free, therefore, to return to some of the most sure and certain principles of religion and traditional virtue—that avarice is a vice, that the exaction of usury is a misdemeanour, and the love of money is detestable, that those walk most truly in the paths of virtue and sane wisdom who take least thought for the morrow. We shall once more value ends above means and prefer the good to the useful. We shall honour those who can teach us how to pluck the hour and the day virtuously as well, the delightful people who are capable of taking direct enjoyment in things, the lilies of the field who toil not, neither do they spin.
>
> But beware! The time for all this is not yet. For at least another hundred years we must pretend to ourselves and to every one that fair is foul and foul is fair; for foul is useful and fair is not. Avarice and usury and precaution must be our gods for a little longer still. For only they can lead us out of the tunnel of economic necessity into daylight.

The new values could not only be premature, they could also be wrong. The year 2000 conditions we have sketched could produce a situation in which illusion, wishful thinking, even obviously irrational behavior could exist to a degree unheard of today. Such irrational and self-indulgent behavior is quite likely in a situation in which an individual is overprotected and has no systematic or objective contact with reality. For example, there are probably many people for whom work is the primary touch with reality. If work is removed, or if important functions are taken from work, the contact these people have with reality will be to some degree impaired. The results—minor or widespread—may become apparent in forms such as political disruption, disturbed families, and personal tragedies—or in the pursuit of some "humanistic" values that many would think of as frivolous or even irrational.

Humanistic values are, of course, a question of definition. While some may judge certain ideologies that invoke humanistic language as better described as sentimental, self-indulgent, or rationalizations of quite irrational feelings of rebelliousness and selfishness, others will accept the ideology. (While this is, of course, more or less a value question, facts and analysis have some relevance to it.)

Consider this question of humanistic versus irrational or indulgent behavior.

In 1926 the British economist Arthur Redford said, in describing the adjustment of British yeomen to industrialization: "In the course of a generation or two it becomes quite 'natural'. . .for a fixed number of hours each day, regulating their exertions constantly. . .there may be some temporary restlessness among the 'hands,' but the routine soon reestablishes itself as part of the ordinary discipline of life." While this may be a rather callous observation, "progress" and other conditions predominantly made the adjustment a necessary one.

In the post affluent, seemingly very secure world of the year 2000, we will not likely, and presumably should not, be willing to ask people to make sacrifices of this order. However, new issues will arise. Consider the following two statements put forth by Berkeley students on signs they were carrying while picketing and later on a BBC television broadcast:

> I am a human being; please do not fold, spindle, or mutilate.
> Life here is a living hell.

One can only agree with the first, assuming we understand precisely in what way the students believe they are not treated as well as IBM cards. Thus it was widely believed, especially in the 1930's and 1940's, by people who thought they were "psychologically sophisticated," that any kind of discipline for children causes undesirable repression, inhibits creativity, and creates neuroses; that almost completely permissive upbringing is necessary for a parent not to "fold, spindle, or mutilate." Today psychoanalysts are emphasizing that a reasonable level of benevolent but firm discipline is very much needed by a child, and that excessive permissiveness is more likely to result in a child marred by guilty wilfulness, irresponsibility, and inadequacy.

Of course, the students would argue that they do not mean anything so extreme, but just that they ought to be treated better than items processed by machines. One can only sympathize with their lack of ability to communicate with a seemingly unfeeling, bureaucratic administration choosing to enforce computer decisions. But to argue that the idiosyncrasies of a computer that allows ten minutes between classes which require fifteen minutes to reach, or that assigns art classes to basements and engineering classes to top-floor rooms with windows, creates difficulties for students, is rather different from arguing that life is a "living hell." The most that students could reasonably say was that the administration made life unnecessarily complicated and frustrating, and had occasionally overstepped its proper bounds. Yet they chose to state (and no doubt felt) these issues in moralistic, politicized, ideological, and emotionally extreme terms. Similarly, increasing numbers of Americans are likely not only to reject currently held work-oriented, achievement-oriented, advancement-oriented attitudes, but are likely to adopt the kind of "spoiled child" attitudes that seem to have characterized at least some of the Berkeley protestors.

I. WHAT IS A STABLE STATE FOR THE ALIENATED-AFFLUENT SOCIETY?

Nevertheless such a society—affluent, humanistic, leisure-oriented, and partly alienated—might be quite stable. It might, in fact, bear some resemblance to some aspects of Greek society (though of course Greek society did not develop primarily because of affluence). We can imagine a situation in which, say, 70 or 80 per cent of people become gentlemen and put a great deal of effort into various types of self-development, though not necessarily the activities which some futurists find most important for a humanistic culture. But one could imagine, for example, a very serious emphasis on sports, on competitive "partner" games (chess, bridge), on music, art, languages, or serious travel, or on the study of science, philosophy, and so on. The crucial point here is that a large majority of the population may feel it important to develop skills, activities, arts, and knowledge to meet very high minimum absolute standards, and a large minority more or less compete to be an elite of elites. One issue is whether or not people who are not well rounded in a number of areas simultaneously, more or less as a gentleman should be, will be considered seriously inferior, or whether it will be sufficient for a person to fulfill himself even if he wishes to do it very narrowly. In both cases, however, there are likely to be at least subtle social pressures for such self-development. In the absence of such social pressures, then, we would still expect much of the same kind of activity but now more in the range of 20 or 30 per cent of the population than 70 or 80 per cent.

Thus there is a very large difference between merely having community acceptance of the right for an individual to spend a lot of time and money on improving himself in this way, and community "demand" that he do so in order to be considered a reasonable or full member of the community, or an educated man. It is hard to believe that in the long run we are not going to get something on the order of the latter in the affluent, postindustrial society. That is, people who are behaving in the new modes will simply look down on those who are not. Indeed there are now such pressures on families in middle-class communities, where there is great emphasis on giving the children dancing lessons, music lessons, and fostering nonutilitarian skills which improve their ability to enjoy themselves and, most important, to be more socially desirable. In other words, middle-class children in the United States are now being treated in a manner not too dissimilar from the way aristocrats treated their children some years ago, except that there is little emphasis on being hard and tough, having a sense of noblesse oblige, and there are somewhat less demanding standards of performance. But while contemporary American parents are in many ways very soft on children, and certainly demand much less in the way of help with chores or housework than they did several generations ago, when it comes to socially important achievements in school, dancing, music, athletics, and so on, they tend to be rather startlingly demanding. While it is true that in many cases the children rather enjoy these activities and do not resent having their schedules so filled, there are many cases in which they do, in fact, feel overburdened by their demanding routine and still feel real pressures to maintain it.

J. A NOTE ON THE PROBLEM OF MAINTENANCE

Many people have been concerned with the possibility that our high future standard of consumption may be negatively affected by the difficulty of maintaining the various vehicles, appliances, and other gadgets involved in personal life, not to speak of the various buildings or personal services. It is clear that there will indeed be a problem. Even today we find that people are throwing out toasters and inexpensive appliances rather than repairing them. Thus, if one compares the United States with a country like Colombia, to take an extreme example, one finds that Americans junk cars, many, many years before they would be junked in Colombia. In Colombia, because there is a shortage of foreign exchange, they do not allow the easy importation of cars. Therefore Colombian mechanics—many of whom are illiterate—have become extraordinarily skilled in keeping old cars running, even manufacturing spare parts. This example gives us a clue to one of the things that might happen. We may import mechanics in the same way we today import resident physicians and interns (in many hospitals 70 to 80 per cent in the United States today come from less-developed nations). One can also easily imagine that as far as moving parts are concerned, the relative cost of repairing could become more and more expensive and the typical solution—both in home and business—may be much like the solution now adopted by the Air Force and many industries: complete replacement of modular units. Thus moving parts such as motors and cleaning elements that are likely to wear out will be built to be maintenance free, and if they fail they will be pulled out and replaced as a unit, the old unit then rebuilt, destroyed, or (as suggested earlier) sent to a less-developed nation for salvage. As for maintenance that cannot be done this way, it is probable that large maintenance organizations will be created that will contract such service for many people (but perhaps not the majority of homeowners, who may prefer to do much of it themselves as a hobby and as a method of "investing" their leisure time). This contract maintenance could be staffed by several kinds of persons. First, there would be Americans who happen to have grown up in a way that leaves them motivated to work, but whose gifts do not enable them to achieve in the more managerial, bureaucratic, intellectual pursuits favored by those who continue to work in this new society. Second, there is likely to be a good deal of import of labor from places like Latin America where competent mechanics will exist who cannot earn very much. (It is easy to imagine that in the year 2000 a mechanic who might be lucky to make two or three thousand dollars a year in Latin American could make ten to twenty thousand a year in the United States.) This import of labor could be via normal immigration, or through special contract labor routes, as in the EEC today.

In view of this potential combination of new designs, new organizations, and useful tasks profitable for an imported class of foreigners, it seems unlikely that the problem of in-place maintenance cannot be solved. For very skilled maintenance, as with expensive watches, one would expect either that repairing will be an extremely valued skill reimbursed proportionately, that there will be imported contract labor, or that the watch or other item will be sent to a foreign country for repair.

K. SOCIAL RESPONSE TO NEW DIFFICULTIES

The most serious issue raised by these speculations (in addition to their validity, of course) is whether they are not just modern manifestations of traditional "aberrant" behavior, or whether they represent a reasonable adjustment or transition state to new traditions and mores. There is also the question of to what degree society will be self-correcting and self-adjusting. Doubtless, however, there will be much room and need for improved social policies. Just as it seems likely that societies have learned to handle routine economic problems sufficiently well to avoid serious depressions, it may be that we have begun to understand social and psychological problems well enough to avoid the partial passivity and failure implicit in these speculations.

While few would now believe that the mere multiplication of productive powers is likely to bring mankind into Utopia, or into anything resembling it, it would be ironic (but not unprecedented) if this multiplication of resources were to create problems too serious for the solutions that those very resources should make feasible. Efforts will doubtlessly be needed to invent and implement ways of coping with the new and unfamiliar problems that will certainly arise. (What these will and should be is beyond the scope of this report, which is intended, of course, to raise such issues rather than to evaluate and settle them.) Yet, despite best efforts, social policies frequently go wrong. . .

WHEN THE MOON IS IN THE SEVENTH HOUSE
AND JUPITER ALIGNS WITH MARS
THEN PEACE WILL GUIDE THE PLANETS
AND LOVE WILL STEER THE STARS
THIS IS THE DAWNING OF THE AGE OF AQUARIUS

—HAIR•

II. The Reign of Automation

If any single concept can symbolize what the Technological Revolution is all about, that concept is automation. The very word conjures up a bewildering mixture of images: a brave new world in which man, liberated from routine drudgery, is free to develop his hitherto latent potentials, or conversely, a dreary wasteland of punchcard conformity, cultural sterility, and technological unemployment. In this section, both of these points of view are ably represented. The first essay is by Marshall McLuhan, whose flamboyant style and optimistic vision of the future have helped to make him one of the leading prophets of the electronic era. In the persepective of history, argues Professor McLuhan, we have passed from the mechanical age, which generated specialization and fragmentation, to the epoch of cybernetics—the new day of electronic media, computers, and automation—which will replace fragmentation with organic unity, which will make man whole and set him free.

93

Marshall McLuhan on Automation •

LEARNING A LIVING

A newspaper headline recently read, "Little Red Schoolhouse Dies When Good Road Built." One-room schools, with all subjects being taught to all grades at the same time, simply dissolve when better transportation permits specialized spaces and specialized teaching. At the extreme of speeded-up movement, however, specialism of space and subject disappears once more. With automation, it is not only jobs that disappear, and complex roles that reappear. Centuries of specialist stress in pedagogy and in the arrangement of data now end with the instantaneous retrieval of information made possible by electricity. Automation is information and it not only ends jobs in the world of work, it ends subjects in the world of learning. It does not end the world of learning. The future of work consists of earning a living in the automation age. This is a familiar pattern in electric technology in general. It ends the old dichotomies between culture and technology, between art and commerce, and between work and leisure. Whereas in the mechanical age of fragmentation leisure had been the absence of work, or mere idleness, the reverse is true in the electric age. As the age of information demands the simultaneous use of all our faculties, we discover that we are most at leisure when we are most intensely involved, very much as with the artists in all ages.

In terms of the industrial age, it can be pointed out that the difference between the previous mechanical age and the new electric age appears in the different kinds of inventories. Since electricity, inventories are made up not so much of goods in storage as of materials in continuous process of transformation at spatially removed sites. For electricity not only gives primacy to *process,* whether in making or in learning, but it makes independent the source of energy from the location of the process. In entertainment media, we speak of this fact as "mass media" because the source of the program and the process of experiencing it are independent in space, yet simultaneous in time. In industry this basic fact causes the scientific revolution that is called "automation" or "cybernation."

•Source: From *UNDERSTANDING MEDIA: The Extensions of Man,* pp. 346-359 by Marshall McLuhan. Copyright © 1964 by Marshall McLuhan. Used with permission of McGraw-Hill Book Company.

94

In education the conventional division of the curriculum into subjects is already as outdated as the medieval trivium and quadrivium after the Renaissance. Any subject taken in depth at once relates to other subjects. Arithmetic in grade three or nine, when taught in terms of number theory, symbolic logic, and cultural history, ceases to be mere practice in problems. Continued in their present patterns of fragmented unrelation, our school curricula will insure a citizenry unable to understand the cybernated world in which they live.

Most scientists are quite aware that since we have acquired some knowledge of electricity it is not possible to speak of atoms as pieces of matter. Again, as more is known about electrical "discharges" and energy, there is less and less tendency to speak of electricity as a thing that "flows" like water through a wire, or is "contained" in a battery. Rather, the tendency is to speak of electricity as painters speak of space; namely, that it is a variable condition that involves the special positions of two or more bodies. There is no longer any tendency to speak of electricity as "contained" in anything. Painters have long known that objects are not contained in space, but that they generate their own spaces. It was the dawning awareness of this in the mathematical world a century ago that enabled Lewis Carroll, the Oxford mathematician, to contrive *Alice in Wonderland,* in which times and spaces are neither uniform nor continuous, as they had seemed to be since the arrival of Renaissance perspective. As for the speed of light, that is merely the speed of total causality.

It is a principal aspect of the electric age that it establishes a global network that has much of the character of our central nervous system. Our central nervous system is not merely an electric network, but it constitutes a single unified field of experience. As biologists point out, the brain is the interacting place where all kinds of impressions and experiences can be exchanged and translated, enabling us to *react to the world as a whole.* Naturally, when electric technology comes into play, the utmost variety and extent of operations in industry and society quickly assume a unified posture. Yet this organic unity of interprocess that electromagnetism inspires in the most diverse and specialized areas and organs of action is quite the opposite of organization in a mechanized society. Mechanization of any process is achieved by fragmentation, beginning with the mechanization of writing by movable types, which has been called the "monofracture of manufacture."

The electric telegraph, when crossed with typography, created the strange new form of the modern newspaper. Any page of the telegraph press is a surrealistic mosaic of bits of "human interest" in vivid interaction. Such was the art form of Chaplin and the early silent movies. Here, too, an extreme speedup of mechanization, an assembly line of still shots on celluloid, led to a strange reversal. The movie mechanism, aided by the electric light, created the illusion of organic form and movement as much as a fixed position had created the illusion of perspective on a flat surface five hundred years before.

The same thing happens less superficially when the electric principle crosses the mechanical lines of industrial organization. Automation retains only as much of the mechanical character as the motorcar kept for the forms of the horse and the carriage. Yet people discuss automation as if we had not passed the oat barrier, and as if the horse-vote at the next poll would sweep away the automation regime.

Automation is not an extension of the mechanical principles of fragmentation and separation of operations. It is rather the invasion of the mechanical world by the instantaneous character of electricity. That is why those involved in automation insist that it is a way of thinking, as much as it is a way of doing. Instant synchronization of numerous operations has ended the old mechanical pattern of setting up operations in lineal sequence. The assembly line has gone the way of the stag line. Nor is it just the lineal and sequential aspect of mechanical analysis that has been erased by the electric speed-up and exact synchronizing of information that is automation.

Automation or cybernation deals with all the units and components of the industrial and marketing process exactly as radio or TV combine the individuals in the audience into new interprocess. The new kind of interrelation in both industry and entertainment is the result of the electric instant speed. Our new electric technology now extends the instant processing of knowledge by interrelation that has long occurred within our central nervous system. It is that same speed that constitutes "organic unity" and ends the mechanical age that had gone into high gear with Gutenberg. Automation brings in real "mass production," not in terms of size, but of an instant inclusive embrace. Such is also the character of "mass media." They are an indication, not of the size of their audiences, but of the fact that everybody becomes involved in them at the same time. Thus commodity industries under automation share the same structural character of the entertainment industries in the degree that both approximate the condition of instant information. Automation affects not just production, but every phase of consumption and marketing; for the comsumer becomes producer in the automation circuit, quite as much as the reader of the mosaic telegraph press makes his own news, or just *is* his own news.

But there is a component in the automation story that is as basic as tactility to the TV image. It is the fact that, in any automatic machine, or galaxy of machines and functions, the generation and transmission of power is quite separate from the work operation that uses the power. The same is true in all servomechanist structures that involve feedback. The source of energy is separate from the process of translation of information, or the applying of knowledge. This is obvious in the telegraph, where the energy and channel are quite independent of whether the written code is French or German. The same separation of power and process obtains in automated industry, or in "cybernation." The electric energy can be applied indifferently and quickly to many kinds of tasks.

Such was never the case in the mechanical systems. The power and the work done were always in direct relation, whether it was hand and hammer, water and wheel, horse and cart, or steam and piston. Electricity brought a strange elasticity in this matter, much as light itself illuminates a total field and does not dictate what shall be done. The same light can make possible a multiplicity of tasks, just as with electric power. Light is a nonspecialist kind of energy or power that is identical with information and knowledge. Such is also the relation of electricity to automation, since both energy and information can be applied in a great variety of ways.

Grasp of this fact is indispensable to the understanding of the electronic age, and of automation in particular. Energy and production now tend to fuse with information and learning. Marketing and comsumption tend to become one with learning, enlightenment, and the intake of information. This is all part of the electric *implosion* that now follows or succeeds the centuries of *explosion* and increasing specialism. The electronic age is literally one of illumination. Just as light is at once energy and information, so electric automation unites production, consumption, and learning in an inextricable process. For this reason, teachers are already the largest employee group in the U.S. economy, and may well become the *only* group.

The very same process of automation that causes a withdrawal of the present work force from industry causes learning itself to become the principal kind of production and consumption. Hence the folly of alarm about unemployment. Paid learning is already becoming both the dominant employment and the source of new wealth in our society. This is the new *role* for men in society, whereas the older mechanistic idea of "jobs," or fragmented tasks and specialist slots for "workers." becomes meaningless under automation.

It has often been said by engineers that, as information levels rise, almost any sort of material can be adapted to any sort of use. This principle is the key to the understanding of electric automation. In the case of electricity, as energy for production becomes independent of the work operation, there is not only the speed that makes for total and organic interplay, but there is, also, the fact that electricity is sheer information that, in actual practice, illuminates all it touches. Any process that approaches instant interrelation of a total field tends to raise itself to the level of conscious awareness, so that computers seem to "think." In fact, they are highly specialized at present, and quite lacking in the full process of interrelation that makes for consciousness. Obviously, they can be made to simulate the process of consciousness, just as our electric global networks now begin to simulate the condition of our central nervous system. But a conscious computer would still be one that was an extension of our consciousness, as a telescope is an extension of our eyes, or as a ventriloquist's dummy is an extension of the ventriloquist.

Automation certainly assumes the servomechanism and the computer. That is to say, it assumes electricity as store and expediter of information. These traits of store, or "memory," and accelerator are the basic features of any medium of communication whatever. In the case of electricity, it is not corporeal substance that is stored or moved, but perception and information. As for technological acceleration, it now approaches the speed of light. All nonelectric media had merely hastened things a bit. The wheel, the road, the ship, the airplane, and even the space rocket are utterly lacking in the character of instant movement. Is it strange, then, that electricity should confer on all previous human organization a completely new character? The very toil of man now becomes a kind of enlightenment. As unfallen Adam in the Garden of Eden was appointed the task of the contemplation and naming of creatures, so with automation. We have now only to name and program a process or a product in order for it to be accomplished. Is it not rather like the case of Al Capp's Schmoos? One had only to look at Schmoo and think longingly of pork chops or caviar, and the Schmoo ecstatically transformed itself into the object of desire. Automation brings us into the world of the Schmoo. The custom-built supplants the mass-produced.

Let us, as the Chinese say, move our chairs closer to the fire and see what we are saying. The electric changes associated with automation have nothing to do with ideologies or social programs. If they had, they could be delayed or controlled. Instead, the technological extension of our central nervous system that we call the electric media began more than a century ago, subliminally. Subliminal have been the effects. Subliminal they remain. At no period in human culture have men understood the psychic mechanisms involved in invention and technology. Today it is the instant speed of electric information that, for the first time, permits easy recognition of the patterns and the formal contours of change and development. The entire world, past and present, now reveals itself to us like a growing plant in an enormously accelerated movie. Electric speed is synonymous with light and with the understanding of causes. So, with the use of electricity in previously mechanized situations, men easily discover causal connections and patterns that were quite unobservable at the slower rates of mechanical change. If we play backward the long development of literacy and printing and their effects on social experience and organization, we can easily see how these forms brought about the high degree of social uniformity and homogeneity of society that is indispensable for mechanical industry. Play them backward, and we get just that shock of unfamiliarity in the familiar that is necessary for the understanding of the life of forms. Electricity compels us to play our mechanical development backward, for it reverses much of that development. Mechanization depends on the breaking up of processes into homogenized but unrelated bits. Electricity unifies these fragments once more because its speed of operation requires a high degree of interdependence among all phases of any operation. It is this electric speed-up and interdependence that has ended the assembly line in industry.

This same need for organic interrelation, brought in by the electric speed of synchronization, now requires us to perform, industry-by-industry, and country-by-country, exactly the same organic interrelating that was first effected in the individual automated unit. Electric speed requires organic structuring of the global economy quite as much as early mechanization by print and by road led to the acceptance of national unity. Let us not forget that nationalism was a mighty invention and revolution that, in the Renaissance, wiped out many of the local regions and loyalties. It was a revolution achieved almost entirely by the speed-up of information by means of uniform movable types. Nationalism cut across most of the traditional power and cultural groupings that had slowly grown up in various regions. Multi-nationalism had long deprived Europe of its economic unity. The Common Market came to it only with the Second War. War is accelerated social change, as an explosion is an accelerated chemical reaction and movement of matter. With electric speeds governing industry and social life, explosion in the sense of crash development becomes normal. On the other hand, the old-fashioned kind of "war" becomes as impracticable as playing hopscotch with bulldozers. Organic interdependence means that disruption of any part of the organism can prove fatal to the whole. Every industry has had to "rethink through" (the awkwardness of this phrase betrays the painfulness of the process), function by function, its place in the economy. But automation forces not only industry and town planners, but government and even education, to come into some relation to social facts.

The various military branches have had to come into line with automation very quickly. The unwieldy mechanical forms of military organization have gone. Small teams of experts have replaced the citizen armies of yesterday even faster than they have taken over the reorganization of industry. Uniformly trained and homogenized citizenry, so long in preparation and so necessary to a mechanized society, is becoming quite a burden and problem to an automated society, for automation and electricity require depth approaches in all fields and at all times. Hence the sudden rejection of standardized goods and scenery and living and education in America since the Second War. It is a switch imposed by electric technology in general, and by the TV image in particular.

Automation was first felt and seen on a large scale in the chemical industries of gas, coal, oil, and metallic ores. The large changes in these operations made possible by electric energy have now, by means of the computer, begun to invade every kind of white-collar and mangement area. Many people, in consequence, have begun to look on the whole of society as a single unified machine for creating wealth. Such has been the normal outlook of the stockbroker, manipulating shares and information with the cooperation of the electric media of press, radio, telephone, and teletype. But the peculiar and abstract manipulation of information as a means of creating wealth is no longer a monopoly of the stockbroker. It is now shared by every engineer and by the entire communications industries. With electricity as energizer and synchronizer, all aspects of production, consumption, and organization become incidental to communications. The very idea of communication as interplay is inherent in the electrical, which combines both energy and information in its intensive manifold.

Anybody who begins to examine the patterns of automation finds that perfecting the individual machine by making it automatic involves "feedback." That means introducing an information loop or circuit, where before there had been merely a one-way flow or mechanical sequence. Feedback is the end of the lineality that came into the Western world with the alphabet and the continuous forms of Euclidean space. Feedback or dialogue between the mechanism and its environment brings a further weaving of individual machines into a galaxy of such machines throughout the entire plant. There follows a still further weaving of individual plants and factories into the entire industrial matrix of materials and services of a culture. Naturally, this last stage encounters the entire world of policy, since to deal with the whole industrial complex as an organic system affects employment, security, education, and politics, demanding full understanding in advance of coming structural change. There is no room for witless assumptions and subliminal factors in such electrical and instant organizations.

As artists began a century ago to construct their works backward, *starting with the effect,* so now with industry and planning. In general, electric speed-up requires complete knowledge of ultimate effects. Mechanical speed-ups, however radical in their reshaping of personal and social life, still were allowed to happen sequentially. Men could, for the most part, get through a normal life span on the basis of a single set of skills. That is not at all the case with electric speed-up. The acquiring of new basic knowledge and skill by senior executives in middle age is one of the most common needs and harrowing facts of electric technology. The senior executives, or "big wheels," as they are archaically and ironically designated, are among the hardest pressed and most persistently harassed groups in human history. Electricity has not only demanded ever deeper knowledge and faster interplay, but has made the harmonizing of production schedules as rigorous as that demanded of the members of a large symphony orchestra. And the satisfactions are just as few for the big executives as for the symphonists, since a player in a big orchestra can hear nothing of the music that reaches the audience. He gets only noise.

The result of electric speed-up in industry at large is the creation of intense sensitivity to the interrelation and interprocess of the whole, so as to call for ever-new types of organization and talent. Viewed from the old perspectives of the machine age, this electric network of plants and processes seems brittle and tight. In fact, it is not mechanical, and it does begin to develop the sensitivity and pliability of the human organism. But it also demands the same varied nutriment and nursing as the animal organism.

With the instant and complex interprocesses of the organic form, automated industry also acquires the power of adaptability to multiple uses. A machine set up for the automatic production of electric bulbs represents a combination of processes that were previously managed by several machines. With a single attendant, it can run as continuously as a tree in its intake and output. But, unlike the tree, it has a built-in system of jigs and fixtures that can be shifted to cause the machine to turn out a whole range of products from radio tubes and glass tumblers to Christmas-tree ornaments. Although an automated plant is almost like a tree in respect to the continuous intake and output, it is a tree that can change from oak to maple to walnut as required. It is part of the automation or electric logic that specialism is no longer limited to just one specialty. The automatic machine may work in a specialist way, but it is not limited to one line. As with our hands and fingers that are capable of many tasks, the automatic unit incorporates a power of adaptation that was quite lacking in the preelectric and mechanical stage of technology. As *any*thing becomes more complex, it becomes less specialized. Man is more complex and less specialized than a dinosaur. The older mechanical operations were designed to be more efficient as they became larger and more specialized. The electric and automated unit, however, is quite otherwise. A new automatic machine for making automobile tailpipes is about the size of two or three office desks. The computer control panel is the size of a lectern. It has in it no dies, no fixtures, no settings of any kind, but rather certain general-purpose things like grippers, benders, and advancers. On this machine, starting with lengths of ordinary pipe, it is possible to make eighty different kinds of tailpipe in succession, as rapidly, as easily, and as cheaply as it is to make eighty of the same kind. And the characteristic of electric automation is all in this direction of return to the general-purpose handicraft flexibility that our hands possess. The programming can now include endless changes of program. It is the electric feedback, or dialogue pattern, of the automatic and computer-programmed "machine" that marks it off from the older mechanical principle of one-way movement.

This computer offers a model that has the characteristics shared by all automation. From the point of intake of materials to the output of the finished product, the operations tend to be independently, as well as interdependently, automatic. The synchronized concert of operations is under the control of gauges and instruments that can be varied from the control-panel boards that are themselves electronic. The material of intake is relatively uniform in shape, size, and chemical properties, as likewise the material of the output. But the processing under these conditions permits use of the highest level of capacity for any needed period. It is, as compared with the older machines, the difference between an oboe in an orchestra and the same tone on an electronic music instrument. With the electronic music instrument, any tone can be made available in any intensity and for any length of time. Note that the older symphony orchestra was, by comparison, a machine of separate instruments that *gave the effect of organic unity.* With the electronic instrument, one *starts* with organic unity as an immediate fact of perfect synchronization. This makes the attempt to create the effect of organic unity quite pointless. Electronic music must seek other goals.

Such is also the harsh logic of industrial automation. All that we had previously achieved mechanically by great exertion and coordination can now be done electrically without effort. Hence the specter of joblessness and property-lessness in the electric age. Wealth and work become information factors, and totally new structures are needed to run a business or relate it to social needs and markets. With the electric technology, the new kinds of instant inter-dependence and interprocess that take over production also enter the market and social organizations. For this reason markets and education designed to cope with the products of servile toil and mechanical production are no longer adequate. Our education has long ago acquired the fragmentary and piecemeal character of mechanism. It is now under increasing pressure to acquire the depth and interrelation that are indispensable in the all-at-once world of electric organization.

Paradoxically, automation makes liberal education mandatory. The electric age of servomechanisms suddenly releases men from the mechanical and specialist servitude of the preceding machine age. As the machine and the motorcar released the horse and projected it onto the plane of entertainment, so does automation with men. We are suddenly threatened with a liberation that taxes our inner resources of self-employment and imaginative participation in society. This would seem to be a fate that calls men to the role of artist in society. It has be effect of making most people realize how much they had come to depend on the fragmentalized and repetitive routines of the mechanical era. Thousands of years ago man, the nomadic food-gatherer, had taken up positional, or relatively sedentary, tasks. He began to specialize. The develop-ment of writing and printing were major stages of that process. They were supremely specialist in separating the roles of knowledge from the roles of action, even though at times it could appear that "the pen is mightier than the sword." But with electricity and automation, the technology of fragmented processes suddenly fused with the human dialogue and the need for over-all consideration of human unity. Men are suddenly nomadic gatherers of knowledge, nomadic as never before, informed as never before, free from fragmentary specialism as never before—but also involved in the total social process as never before; since with electricity we extend our central nervous system globally, instantly interrelating every human experience. Long accustomed to such a state in stock-market news or front-page sensations, we can grasp the meaning of this new dimension more readily when it is pointed out that it is possible to "fly" unbuilt airplanes on computers. The specifications of a plane can be programmed and the plane tested under a variety of extreme conditions before it has left the drafting board. So with new products and new organizations of many kinds. We can now, by computer, deal with complex social needs with the same architectural certainty that we previously attempted in private housing. Industry as a whole has become the unit of reckoning, and so with society, politics, and education as wholes.

Electric means of storing and moving information with speed and precision make the largest units quite as manageable as small ones. Thus the automation of a plant or of an entire industry offers a small model of the changes that must occur in society from the same electric technology. Total interdependence is the starting fact. Nevertheless, the range of choice in design, stress, and goal within that total field of electromagnetic interprocess is very much greater than it ever could have been under mechanization.

Since electric energy is independent of the place or kind of work-operation, it creates patterns of decentralism and diversity in the work to be done. This is a logic that appears plainly enough in the difference between firelight and electric light, for example. Persons grouped around a fire or candle for warmth or light are less able to pursue independent thoughts, or even tasks, than people supplied with electric light. In the same way, the social and educational patterns latent in automation are those of self-employment and artistic autonomy. Panic about automation as a threat of uniformity on a world scale is the projection into the future of mechanical standardization and specialism, which are now past.

With great enthusiasm, Marshall McLuhan has proclaimed that the age of electronics and cybernetics will liberate man from drudgery and enable him to be truly human. Yet, many thoughtful observers have looked upon the Technological Revolution as a highly dangerous phenomenon, bearing tragic implications for the destiny of mankind. One of the most distinguished critics of the de-humanizing tendencies of science and technology is Lewis Mumford, author of some 20 works, social philosopher, educator, expert on city planning, architecture, history, and culture generally. The following selections are taken from the most recent of a series of four volumes in which Mumford has surveyed the whole of human history and prehistory in order to explain the origin and signifi-cance—and the inherent perils—of technology.

Lewis Mumford on the Evils of Automation*

*   *   *   *   *

3. THE REMOVAL OF LIMITS

Every earlier system of production, whether in agriculture or in handicraft, developed in response to human needs and was dependent upon the energy derived mainly from plant growth, supplemented by animal, wind, and water power. His productivity was restricted, not merely by available natural resources and human capacity, but by the variety of non-utilitarian demands that accompanied it. Esthetic design and qualitative excellence took precedence over mere quantitative output, and kept quantification within tolerable human limits.

In the mechanized, high-energy system developed during the last two centuries, these conditions have been radically altered; and one of the results of

•Source: From THE MYTH OF THE MACHINE: THE PENTAGON OF POWER, pp. 172-185, 274-275 copyright © 1964, 1970 by Lewis Mumford. Reprinted by permission of Harcourt Brace Jovanovich, Inc. (Sections have been omitted with permission of the publisher.)

commanding a plethora of energy is to place the stress on precisely those parts of our technology that demand the largest quantities of it; namely, those that make the fullest use of power-machines. This new industrial complex is based upon a group of postulates so self-evident to those who have produced the system that they are rarely criticized or challenged—indeed almost never examined—for they are completely identified with the new 'way of life.' Let me list these postulates once more, though I have already touched on them in examining the mechanical world picture.

First: man has only one all-important mission in life: to conquer nature. By conquering nature the technocrat means, in abstract terms, commanding time and space; and in more concrete terms, speeding up every natural process, hastening growth, quickening the pace of transportation, and breaking down communication distances by either mechanical or electronic means. To conquer nature is in effect to remove all natural barriers and human norms and to substitute artificial, fabricated equivalents for natural processes: to replace the immense variety of resources offered by nature by more uniform, constantly available products spewed forth by the machine.

From these general postulates a series of subsidiary ones are derived: there is only one efficient speed, *faster;* only one attractive destination, *farther away;* only one desirable size, *bigger;* only one rational quanitative goal, *more.* On these assumptions the object of human life, and therefore of the entire productive mechanism, is to remove limits, to hasten the pace of change, to smooth out seasonal rhythms and reduce regional contrasts—in fine, to promote mechanical novelty and destroy organic continuity. Cultural accumulation and stability thus become stigmatized as signs of human backwardness and insufficiency. By the same token, any institution or way of life, any system of education or production that imposes limits, retards change, or converts the imperious will to conquer nature into a relation of mutual aid and rational accommodation, threatens to undermine the power-pentagon and the scheme of life derived from it.

Now this supposed necessity to conquer nature is not quite so innocent in either its origins or its intentions as might seem. In part, at least, it applies unscrupulously to nature the more ancient ambitions of military conquest and imperialist exploitation; but in part, unfortunately, it is also due to a profound fault in Christian theology, which regarded the earth as man's exclusive property, designed by God solely for his use and enjoyment, and further looked upon all other living creatures as without souls, and so subject to the same treatment as inanimate things. (The present turning of the young to Hindu and Buddhist conceptions may be hopefully interpreted as an attempt to overcome this original ecological error. For the meek and the humble, not the proud, alone are fit to inherit the earth.)

Because these traditional attitudes toward man and nature supported the dominant power motives in post-medieval society, the new system of production lacked any method for normalizing wants or controlling quantity: it not merely lacked them but purposely broke down any older methods such as a concern with fine workmanship or esthetic expression.

Thanks to the proficiency of the machine, the problem of older societies, that of scarcity and insufficiency, was—at least in theory—solved; but a new problem, equally serious but at just the opposite extreme, was raised: the problem of quantity. This problem has many aspects: not merely how to distribute the potential abundance of goods justly, so that the whole community will benefit, but how to allocate the investment in machine-centered organizations without negating or destroying those many human activities and functions that are injured rather than helped by automation. The first of these problems has been far more successfully dealt with in many primitive communities than under any industrialized regime.

The bitter reproach that became popular in America during the economic depression of the nineteen-thirties, "starvation in the midst of plenty," reflected the breakdown in a distribution system whose conventions were based on scarcity. But an equally vexatious form of starvation is that which has been caused through the introduction of mechanized habits of life and automatic machine, by the pressure of overwhelming abundance. One might call this the Strasbourg-goose syndrome: gorging or forced feeding for the sake of further fattening a system of automation that produced quantities beyond the normal requirements of consumption.

Though I must postpone a more comprehensive discussion of this problem to a later point, this is the place to examine the impact of automation in a society that takes quantification and material expansion to be an ultimate good. And since the condition to be analyzed now exists in almost every phase of automation, from food production to nuclear weapons, I shall confine myself largely to the field I have the closest acquaintance with: the automation of knowledge. In this area conventional mechanical automation has up to now played only a small part.

As has happened again and again in technics, the critical step that led to general automation took place in the organization of knowledge before any appropriate automatic machinery was invented. The process has been dated and explained, stage by stage, by an historian of science, Derek Price, in 'Science Since Babylon,' and condensed, with certain necessary corrections, in a later essay.

Well before the automatic machines of the nineteenth century had been invented, science had perfected within its own realm a system of subdivided labor, operating with the standardized parts, confined to limited motions and processes, which paralleled in efficiency Adam Smith's favorite example of pin-making.

The means of effecting this immense outpouring of standardized knowledge, Price points out, was a new method of multiplying and communicating scientific information by means of a small standard unit, the scientific paper, whereby reports on isolated observations and experiments could be promptly circulated in scientific journals. This practical device, based on the earlier invention of the printing press, proved the effective starting point for the systematic automation of knowledge. By now the productivity in this area rivals anything that has been achieved in industrial manufacture. Periodical publication is in itself a phase of automation: once a periodical is set up, the regular flow of material and its regular publication is no longer subject to spontaneous fluctuations of supply or erratic publishing demands: the process instigates the product and punctuates the result—automatically.

Observe the interplay between the mass production of goods and the mass publication of scientific knowledge. Beginning with a single scientific journal in 1665, Price tells us that there were a hundred at the beginning of the nineteenth century, a thousand by the middle, and ten thousand by 1900. We are already on the way to achieve 100,000 journals in another century. Even allowing for the great increase in population, this is a gigantic advance. In the meanwhile, the enormous output of duplicating machines of every kind, from the mimeograph to the microfilm and Xerox, has multiplied the product. And here again the result is typical of the entire system: before any part of this process, except large-scale printing, was mechanically automated, the entire system exhibited all the virtues and defects of any completely automated unit—expanding productivity in quantities that are unassimilable, without reintroducing the human selections and abstentions that have been excluded from the system.

4. THE TRIUMPH OF AUTOMATION

The place to appraise the whole process of mechanization and mass production is at the terminal point already visible in many areas: total automation. Now neither the idea of automation, nor the process itself, belongs exclusively to the modern age: nor, more importantly, was either aspect dependent solely upon mechanical inventions. Growing plants are natural agents that automatically turn the sun's energy into leafy tissue; and the synthetic reproduction of this process in an automated chemical plant would not make it in any degree more automatic. So, too, the gravity flow system of conveying water through a pipe from a mountain spring, as in the ancient Place at Knossos, was quite as automatic and efficient—and even more reliable than—the operation of an electrically driven hydraulic pump today. When Aristotle used the term automation, it was to describe those natural changes that take place, as in a chemical reaction, without any final purpose. But long before man had any scientific grasp of the role of organic automatism within the body the idea itself had taken hold of his mind; and it was from the first associated with three magical aims; super-human power, material abundance, and remote control.

Central to these magic aspirations was, for obvious reasons, material abundance; this proved indeed to be the tempting immediate bait that concealed the collective trap of external power and centralized control. As early as 446 B.C. the Greek poet Telecleides, himself probably echoing many unrecorded fables, pictured the Golden Age as one when "the earth bore neither fear nor disease, but all things appeared of their own accord; for every stream flowed with wine and barley cakes fought with wheat cakes to enter the mouths of men." Though the machine plays no part in this magical wish, the fantasy dwells on those gustatory joys and that effortless existence which people still associate with automation. As for the life so pictured, it was nothing less than the kind of existence that kings, nobles, and rich magnates had long enjoyed.

With this promise of abundance went another persistent wish: the idea of finding a mechanical substitute to take over the burden of painful human toil. Though Babylonian legends picture the gods as creating man purely for the purpose of performing back-breaking tasks for them, the more self-confident Greeks pictured their blacksmith god, Hephaistos, as proving his skill by creating a lifelike bronze automaton—historically the first of a long line of spectral robots that still haunt the minds of modern engineers.

In the very act of affirming the necessity of slavery by dismissing the idea that self-acting machines for weaving or building might be invented, Aristotle showed that the possibility of manufacturing automatons was active in the Greek mind: so we need not be surprised that Heron of Alexandria, a few centuries later, described a more elaborate automaton, that of a naval dockyard, where puppets cut and sawed timber. Here in playful form was the earliest small-scale model of the automated factory.

Since the fantasies of automation and absolute power have historically gone together, it is hardly surprising that absolute monarchs in all ages have persistently delighted in automatons, as symbolic witnesses to the unqualified power they themselves sought to exercise. Marco Polo, happily, has transmitted to us the boast of the Great Khan who regarded Christians as "ignorant, inefficient persons" because they did not possess the faculty of performing anything miraculous, whereas, he pointed out, "when I sit at table, the cups that were in the middle of the hall come to me, filled with wine and other beverages, spontaneously, and without being controlled by human hand." This technical facility, Kublai Khan plainly indicated, was proof of his own power and total control. He even went so far, in the same breath, as to anticipate the further extensions of this claim, made by scientists in our own day; for he boasted that his magicians had the power of controlling bad weather, and obliging it to retire to any quarter of the heavens. Marco Polo, unfortunately, neglected to verify this claim.

None of these motives was absent from the later developments of mechanization: but if ages passed before they became realizable, it was because these deep subjective impulses could not be harnessed until the necessary mechanical components were invented. Slaves and servants, treated as if they were such mechanical parts, may actually have delayed the coming of automation, for even now, it has been found, human organisms are still the best available all-round servomechanisms, cheaper to produce, easier to keep in order, more responsive to signals, than the most finicking mechanical robot.

Once more we come back to the mechanical clock. Apart from feedback the invention and perfection of the clock was the decisive move toward automation; for it provided the master model for many other automatic machines; and it reached a degree of perfection finally, in the eighteenth-century chronometer, which set a standard for other technological refinements. The one element lacking in the clock until the electric clock was invented, an automatic source of power, was provided at an early stage for coarser uses by the watermill: the automatic mine-pumping apparatus shown by Agricola in 'De Re Metallica,' and the equally automatic silkreeling machine with multiple spools, illustrated by Zonca in 1607 in his 'Novo Teatro dei Machine e Edifici,' were but the late instances of a series of earlier automatic machines that lacked only a cybernetic regulator of the process and the output to become completely automated. Those who still imagine that automation first took place in the nineteen-forties, and was impossible until the computer was invented, have a lot of homework to do.

Purely as a machine, the clock remained—until the computer—the equal of all other automatic machines in refinement of construction and accuracy of operation; and long before this further improvement took place in any other area, the reduction of the fifteenth-century clock, with its clumsy clanking works, to the small, portable watch set a goal for later forms of miniaturization. What was lacking until the seventeenth century then, was not automatons but a fully developed *system* of automation, and this awaited two things: the construction of the new mechanical world picture and an increase of demand sufficient to justify the installation of expensive prime movers and batteries of elaborate machines, kept in constant use. The sporadic need, the fitful demand, the special adaptation to regional resources or personal desires—all characteristic of small communities and handicraft operations—offered no incentive to complete automation. Rather, they remained obstacles to its achievement.

Here we come to the great paradox of both early mechanization and its ultimate expression in automation: so far from being responses to a mass demand, the enterpriser had in fact to create it; and in order to justify the heavy capital investment necessary to create automatic machines and automatic factories that assembled these machines in larger working units, it was necessary to invade distant markets, to standardize taste and buying habits, to destroy alternative choices, and to wipe out competition from smaller industrial competitors, more dependent upon intimate face-to-face relations and more flexible in meeting consumer demands.

Sigfried Giedion's classic analysis of the processes of rationalization and automation, in 'Mechanization Takes Command,' demonstrates that the result of automation is not necessarily a better product; it merely enables the same product to be sold at a larger profit in a mass market. The growth of automated breadmaking has driven thousands of local bakers out ot existence; but the result is neither a cheaper nor a superior loaf. What automation has done is to funnel its local energy-economies into long-distance-transportation, advertising, higher salaries and profits, and further investments in plant expansion to the same ends. *The desired reward of this magic is not just abundance but absolute control.* Where the industry is sufficiently well unionized, the result is to extend this system of mass control to the labor union itself, under pseudodemocratic self-government.

5. THE SAND IN THE WORKS

The process of automation has gone on steadily during the last century and a half. In the first stages of mechanization the number of operations performed by any one laborer was likewise lessened, with a consequent loss of intelligent participation, as well as initiative, in the process as a whole. But the success of mechanization was gauged in terms of lessening the ratio of man-hours to the units of production, until finally, with complete automation and cybernetic control, only the minimum supervision of the whole plant remained, while the "work" left was little more than inspection and repair. Though computers and cybernetic controls are necessary when the overall unit is a complex assembly, there are essential likenesses between the automatic loom and an electric computer. For the latter, too, requires a human being to design it, to program it, and to monitor it.

When human monitors are lacking, serious breakdowns may occur—as more than one comic incident testifies. Witness the case of a defective machine in a fully automated English nuclear plant, which was programmed to call automatically for help in such case to a London station. Unfortunately, the taped voice which said: "Send an engineer at once" was answered by an equally automated telephone, which replied: "The number you are calling has been changed to . . . ," giving the new number. But the calling system had not been programmed to deal with new numbers, so getting no proper answer it kept on insistently dialing the original number—until the prolonged breakdown tardily awakened a human mind capable of intervening and summoning help.

But it is not to point out their frailties in operation that I have traced the tendency of mechanization and automation to form a self-enclosed system: one must expect residual errors or malfunctions in any product that comes from the hand of man; and where the object is an appropriate one the gains from automation may far outweigh the occasional disabilities one encounters. The point is that the most massive defects of automation are those that arise, not from failures, but from its indisputable triumphs, above all, in those departments where the most optimistic hopes and boasts have been completely justified.

Let me emphasize: work in all its aspects has played a decisive, formative part in the enlargement of man's mind and the enrichment of his culture, not becuase man is identifiable solely as a tool-using animal, but because work is one of the many activities that have stimulated his intelligence and enlarged his bodily capacities. But if, for argument's sake, one accepts the still lingering anthropological identification of man's basic nature with tool-using and tool-making, what then should one say about the cumulative results of mechanization and automation, as they affect man's adaptive intelligence?

What merit is there in an over-developed technology which isolates the whole man from the work-process, reducing him to a cunning hand, a load-bearing back, or a magnifying eye, and then finally excluding him altogether from the process unless he is one of the experts who designs and assembles or programs the automatic machine? What meaning has a man's life as a worker if he ends up as a cheap servo-mechanism, trained solely to report defects or correct failures in a mechanism otherwise superior to him? If the first step in mechanization five thousand years ago was to reduce the worker to a docile and obedient drudge, the final stage automation promises today is to create a self-sufficient mechanical electronic complex that has no need even for such servile nonentities.

Curiously, all the while automatic processes were being perfected in industry, the leaders of nineteenth-century thought stressed, as never before, the human value of work as a way of easing anxiety and increasing the sum total of human happiness. Such a recognition of the dignity and value of work had been going on, sporadically, for a long time. While the pride of craft was an old one, it had been re-enforced by the creed of the Benedictine order, which held that 'to labor is to pray'; and it had gained institutional support in the medieval guild, which made a whole network of social relationships center in the workshop and its fellowship. Thus work in all its forms came to be regarded as the central activity of life: was it not indeed on these grounds that both manufacturers and workers despised the idle, playful, landed aristocracy, who for lack of serious work turned fox hunting and grouse shooting, polo and war and amatory adventure, into substitute forms of work, just as active, just as demanding?

Surely the time has come to reconsider the abolition of work. If work has been an integral part of human culture, and thus one of the active determinants of man's own nature for at least half a million years—and had perhaps its dim beginnings a million and a half years earlier, in the little hominoid ape that many anthropologists have too hastily identified as 'man'—what will remain of man's life if these formative activities are wiped out by universal cybernetics and automation?

Strange to say, it is only recently that the full implications of such a blanking out of the largest portion of man's working life has presented itself as a problem, though automation has been steadily gaining ground. Even now only a few realize that this problem, once honestly stated, seriously calls into question the ultimate goals of automation. As for the eventual assemblage of a completely automated world society, only innocents could contemplate such a goal as the highest possible culmination of human evolution. It would be a final solution to the problems of mankind, only in the sense that Hitler's extermination program was a final solution for the 'Jewish problem.'

6. THE PARADOX OF AUTOMATION

Here we face the great paradox of automation, put once and for all in Goethe's fable of the Sorcerer's Apprentice. Our civilization has cleverly found a magic formula for setting both industrial and academic brooms and pails of water to work by themselves, in ever-increasing quantities at an ever-increasing speed. But we have lost the Master Magician's spell for altering the tempo of this process, or halting it when it ceases to serve human functions and purposes, though this formula (foresight and feedback) is written plainly on every organic process.

As a result we are already, like the apprentice, beginning to drown in the flood. The moral should be plain: unless one has the power to stop an automatic process—and if necessary reverse it—one had better not start it. To spare ourselves humiliation over our failure to control automation, many of us now pretend that the process conforms exactly to our purposes and alone meets all our needs—or, to speak more accurately, we cast away those qualifying human traits that would impede the process. And as our knowledge of isolatable segments and fragments becomes infinitely refined and microscopic, our ability to interrelate the parts and to bring them to a focus in rational activities continues to disappear.

In even the most restricted area of knowledge—let us say virus diseases in the gastrointestinal tract of elderly earthworms—it is difficult for the most conscientious scholar to keep his head above water. To cope with the tidal wave of rapidly processed knowledge, the academic world has now taken the final step toward total automation: it has resorted to further 'mechanical' agents that only aggravate the original condition, because they seek to deal only with the result and do not dream of attacking the causes—namely, their own preconceptions and methods. The exponents of mass production of knowledge have created a hundred journals devoted only to abstracts of papers; and now a further abstract of all these abstracts has been proposed. At the terminal stage of this particular solution, all that will be left of the original scientific or scholarly paper will be a little vague noise, at most a title and a date, to indicate that someone has done something somewhere—no one knows what and Heaven knows why.

Though this program for the automatic mass production of knowledge originated in science, and shows characteristic seventeenth-century limitations, it has been imitated in the humanities, particularly in American universities, as a sort of status symbol, to underwrite budget requests in competition with the physical and social sciences, and to provide a quantitative measure for professional promotions. Whatever the original breach between the sciences and the humanities, in method they have now—*pace* Charles Snow!—become one. Though they run different assembly lines, they belong to the same factory. The mark of their common deficiency is that neither has given any serious consideration to the results of their uncontrolled automation.

Even a generation ago there was still a large margin for free activity and independent thinking within higher education. But today most of our larger academic institutions are as thoroughly automated as a steel-rolling mill or a telephone system: the mass production of scholarly papers, discoveries, inventions, patents, students, Ph.D's, professors, and publicity—not least, publicity!—goes on at a comparable rate; and only those who identify themselves with the goals of the power system, however humanly absurd, are in line for promotion, for big research grants, for the political power and the financial rewards allotted to those who 'go with' the system. The voluminous flow of corporate capital into the Educational Establishment, with a corresponding rise in money incentives for research, has proved in the United States the final step in making the University an integral part of the new power system.

Meanwhile, a vast amount of valuable knowledge becomes relegated, along with an even greater amount of triviality and trash, to a mountainous rubbish heap. For lack of a method with built-in qualitative standards, fostering constant evaluation and selectivity, and with assimilative processes that, as in the digestive system, would control both appetite and feeding, the superficial order of the individual packet is offset by the nature of the end product: for to know more and more about less and less is in the end simply to know less and less.

As a means of creating an orderly and intelligible world, the automation of knowledge has already come close to total bankruptcy; and the current revolt of the university students, along with the even more threatening regression to total nihilism, is a symptom of this bankruptcy.

Do not, I beg, misinterpret this factual description of the automation of knowledge as mischievous satire on my part; still less must it be taken as an attack on science, scholarship, or the many exquisite feats of electronic and cybernetic technology. No one but an idiot would belittle the immense practical benefits and the exhilarating prospects for the human spirit that the sciences, abetted by technics, have opened up. All I am saying here is that the 'automation of automation' is now a demonstrable irrationality in every department where it has taken hold: in the sciences and humanities as much as in industry and warfare. And I suggest that this is an inherent defect of any completely automated system, not an accidental one.

This irrationality was humorously summarized, with feigned exactitude, by Derek Price; for he calculated that at the present rate of acceleration in scientific productivity alone, within a couple of centuries there will be dozens of hypothetic scientists for every man, woman, child, and dog on the planet. Fortunately, ecology teaches us that under such conditions of overcrowding and stress, most of the population will have died off before it reaches this point.

But there is no need to wait for the ultimate breakdown of this system to foresee its consequences. Long before nearing the theoretic end, the symptoms have become ominous. Already the great national and university libraries are at their wits' end, not merely to find place for the books already acquired— selective though that process has always been—but even to catalog promptly the annual output of books, papers, and periodicals. Without pausing to weigh the consequences many administrators are now playing with the desperate notion of abandoning the preservation of books entirely, as an obsolete form of the permanent record, and transferring the contents at once to microfilms and computers.

Unfortunately, "information retrieving," however swift, is no substitute for discovering by direct personal inspection knowledge whose very existence one had possibly never been aware of, and following it at one's own pace through the further ramifications of relevant literature. But even if books are not abandoned, but continue their present rate of production, the multiplication of microfilms actually magnifies the central problem—that of coping with quantity—and postpones the real solution, which must be conceived on quite other than purely mechanical lines: namely, by a reassertion of human selectivity and moral self-discipline, leading to more continent productivity. Without such self-imposed restraints the over-production of books will bring about a state of intellectual enervation and depletion hardly to be distinguished from massive ignorance.

As the quantity of information increases in every field, to a point where it defies individual appraisal and assimilation, an ever larger part of it must be channeled through official distribution agencies. Though a trickle of fresh or unorthodox knowledge may still filter through to a miniscule minority by means of print, nothing will be transmitted further that does not conform to the current standards of the megamachine. This was neatly illustrated during the mounting Vietnam crisis in the United States, when television gave equal time to speakers who favored the official policy of seeking military victory and to those who favored entering into negotiations; but sedulously refrained from inviting those who, like myself, had put the case for unconditional withdrawal of American forces—at a time when this could still have been done without confessing a humiliating defeat.

Both the ancient and the contemporary control systems are based, essentially, on one-way communication, centrally organized. In face-to-face communication even the most ignorant person can answer back, and he has various means at his command besides the word—the expression of his face, the stance of his body, even threat of bodily assault. As the channels of instantaneous communication become more elaborate, the response must be officially staged, and this means, in ordinary circumstances, externally controlled. The attempt to overcome this difficulty with 'opinion polls' is only a more insidious way of maintaining control. The more complex the apparatus of transmission, the more effectively does it filter out every message that challenges or attacks the Pentagon of Power.

Though total control over the media of communication seems to give the modern megamachine a great advantage over the crude, earlier model, it is likely that its expansion may in the end hasten its breakdown, because of a lack of information needed for its own efficient performance. The refusal to accept such information even when offered becomes more ingrained as the system itself becomes more closely knit together.

Today the increasing number of mass protests, sit-downs, and riots—physical acts rather than words—may be interpreted as an attempt to break through the automatic insulation of the megamachine, with its tendency to cover up its own errors, to refuse unwelcome messages, or to block transmission of information damaging to the system itself. Smashed windows, burning buildings, broken heads are means of making humanly important messages take possession of the unmindful medium and so resume, though in the crudest form possible, two-way communication and reciprocal intercourse.

Once automatic control is installed one cannot refuse to accept its instructions, or insert new ones, for theoretically the machine cannot allow anyone to deviate from its own perfect standards. And this bring us at once to the most radical defect in every automated system: for its smooth operation this under-dimensioned system requires equally under-dimensioned men, whose values are those needed for the operation and the continued expansion of the system itself. The minds that are so conditioned are incapable of imagining any alternatives. Having opted for automation, they are committed to flouting any subjective reaction and to wiping out human autonomy—or indeed any organic process that does not accept the system's peculiar limitations.

Here, at the core of automation, lies its principal weakness once the system becomes universal. Its exponents, even if they are able to recognize its deficiencies, see no way of overcoming them except by a further extension of automation and cybernation. Thus a large-scale processing of compulsory leisure is now in order, to find profit-making substitutes for the vanished pleasures of work, which once brought an immediate human reward within the workshop, the marketplace, or the farm, both through the job itself and the many occasions for human association it opened up. The fact is, however, that an automatic system as a whole, once established, can accept no human feedback that calls for a cutback: therefore it accepts no evaluation of its deleterious results, still less is it ready to admit the need for correcting its postulates. Quantity is all. To question the value of mere quantitative increase in terms of its contribution to human well-being is to commit heresy and weaken the system.

Here finally we face another difficulty derived from automation itself. As the mechanical facilities of our educational institutions expand, with their heavy investment in nuclear reactors, their computers, their TV sets and tape recorders and learning machines, their machine-marked 'yes-or-no' examination papers, the human contents necessarily shrink in significance. What automation has done in every department where it has taken full command is to make difficult—in many cases impossible—the give-and-take that has existed hitherto between human beings and their environment; for the constant dialogue that is so necessary for self-knowledge, for social cooperation, and for moral evaluation and rectification, has no place in an automated regimen.

When Job's life miscarried, he was able, at least in imagination, to confront God and criticize his ways. But the suppression of personality is already so complete in an automated economy that the reputed heads of our great organizations are as incapable of changing its goals as the lowliest filing clerk. It is the system itself that, once set up, gives orders. As for anyone's confronting the principals in person, our automatic agencies are as obscure and as bafflingly inaccessible as the authorities that Franz Kafka pictured in his accurate prophetic nightmare, 'The Trial.' Humanly speaking, then, the proper name for automation is self-inflicted impotence. That is the other side of 'total control.'

While our technicians have been designing machines and automated systems to take on more of the attributes of living organisms, modern man himself, to fit into this scheme, finds he must accept the limitations of the machine and not ask for those qualitative and subjective attributes which the Mechanical World Picture originally failed to acknowledge, and which the machine-process, inevitably, does not possess.

What has proved quite as serious is that as the system of automation becomes more highly articulated, and thereby more self-sufficient and self-enclosed, it is less possible for anyone to intervene in the process, to alter its pace, to change its direction, to limit its further extension, or to reorient its goal. The parts may be flexible and responsive, as individual computers that play chess have demonstrated: but the large automated system becomes increasingly rigid. Automation has thus a qualitative defect that springs directly from its quantitative accomplishments: briefly it increases probability and decreases possibility. Though the individual component of an automatic system may be programmed like a punch card on a motor-car assembly line, to deal with variety, the system itself is fixed and inflexible: so much so that it is little more than a neat mechanical model of a compulsion neurosis, and perhaps even springs from the same ultimate source—anxiety and insecurity.

*   *   *   *   *

3. THE ALL-SEEING EYE

In Egyptian theology, the most singular organ of the Sun God, Re, was the eye: for the Eye of Re had an independent existence and played a creative and directive part in all cosmic and human activities. The computer turns out to be the Eye of the reinstated Sun God, that is, the Eye of the Megamachine, serving as its 'Private Eye' or Detective, as well and the omnipresent Executive eye, he who exacts absolute conformity to his commands, because no secret can be hidden from him, and no disobedience can go unpunished.

The principal means needed to operate the megamachine correctly and efficiently were a concentration of power, political and economic, instantaneous communication, rapid transportation, and a system of information storage capable of keeping track of every event within the province of the Divine King: once these accessories were available, the central establishment would also have a monopoly of both energy and knowledge. No such complete assemblage had been available to the rulers of the pre-scientific ages: transportation was slow, communication over a distance remained erratic, confined to written messages carried by human messengers, while information storage, apart from tax records and books, was sporadic and subject to fire and military assault. With each successive king, essential parts would require reconstruction or replacement. Only in Heaven could there exist the all-knowing, all-seeing, all-powerful, omnipresent gods who truly commanded the system.

With nuclear energy, electric communication, and the computer, all the necessary components of a modernized megamachine at last became available: 'Heaven' had at last been brought near. Theoretically, at the present moment, and actually soon in the future, God—that is, the Computer—will be able to find, to locate, and to address instantly, by voice and image, via the priesthood, any individual on the planet: exercising control over every detail of the subject's daily life by commanding a dossier which would include his parentage and birth; his complete educational record; an account of his illnesses and his mental breakdowns, if treated; his marriage; his sperm bank account; his income, loans, security payments; his taxes and pensions; and finally the disposition of such further organs as may be surgically extracted from him just prior to the moment of his official death.

In the end, no action, no conversation, and possibly in time no dream or thought would escape the wakeful and relentless eye of this deity: every manifestation of life would be processed into the computer and brought under its all-pervading system of control. This would mean, not just the invasion of privacy, but the total destruction of autonomy: indeed the dissolution of the human soul.

Half a century ago, the foregoing description would have seemed too crude and overwrought to be accepted even as satire: H. G. Wells' 'Modern Utopia,' which tentatively provided for a central identification system, did not dare carry the method through into every detail of life. Even twenty years ago, only the first faint outlines of this modern version of the Eye of Re could be detected by such a prescient mind as that of Norbert Wiener. But today the grim outlines of the whole system have been laid down, with the corroborative evidence, by a legal observer, Alan F. Westin, as an incidental feature of a survey of the numerous public agencies and technological devices that are now encroaching on the domain of private freedom.

What Westin demonstrates, also in passing, is that the countless record files, compiled by individual bureaucracies for their special purposes, can already be assembled in a single central computer, thanks to the fantastic technological progress made through electro-chemical miniaturization: not merely the few I have just picked out, but civil defense records, loyalty security clearance records, land and housing records, licensing applications, trade union cards, social security records, passports, criminal records, automobile registration, driver's licenses, telephone records, church records, job records—indeed the list becomes finally as large as life—at least of abstracted, symbolically attenuated, recordable life.

The means for such total monitoring came about through the quantum jump from macro- to micro-mechanics: so that the seemingly compact microfilm of earlier decades, to quote Westin's words, "has now given way to photochromatic microimages that make it possible to reproduce the complete bible on a thin sheet of plastic less than two inches square, or to store page by page copies of all books in the library of Congress in six four-drawer filing cabinets." The ironic fact that this truly colossal leap was a product of research by the National Cash Register Company does not detract from the miraculous nature of this invention. . .

"...in the American psychology, the city has been a basically suspect institution, reeking of the corruption of Europe, totally lacking that sense of spaciousness and innocence of the frontier and the rural landscape."

John Lindsay●

III. Megalopolis: The Present and Future of the Super-City

One cause of the urban crisis besetting America today has been the social impact of technology upon the South. With the mechanization of cotton-picking and the strip-mining of coal, vast numbers of poor whites and poor Blacks found themselves forced out of their traditional modes of life and employment. In vast numbers, they poured into the industrial cities of the North—the result being a series of crises in race relations, welfare, housing, sanitation, police functions, and education. In fact, to many observers, the contemporary American city seemed to consist mainly of ghettos, crime, pollution, corruption, and perpetual crisis. In the following selection, Lewis "Studs" Terkel, prize-winning radio broadcaster, author, and interviewer uses his taperecorder—one more product of the Technological Revolution—to capture the immediacy and the actuality of life in the great Chicago ghetto. He has given his interviewee the pseudonym "Barbara Hayes."

●Source: From THE CITY by John V. Lindsay. Copyright © 1969, 1970 by W. W. Norton & Company, Inc. With the permission of the publisher.

Studs Terkel on
the Chicago Ghetto •

BARBARA HAYES, Late Thirties [display]

Number 1510: a doorway facing out on the gallery. (Gallery: the official name for the balcony that extends out on each story of the building. The words *porch* and *gallery* are used interchangeably.) An apartment in the world's largest public-housing project: the Robert Taylor Homes, on south State Street, extending from Pershing Road (39th Street) to 54th. We're in the kitchen of a six-room flat. Somewhere in the vague distance a child is whimpering. "The little one," murmurs Mrs. Hayes. "Her afternoon nap." A small boy of four or five is observing us.

She has eight children, four boys and four girls—including a set of twins—ranging in age from sixteen to the baby. She has recently been separated from her husband and is on ADC. She came to Chicago in 1954, and all these years she has lived in a project.

She attended Southern Illinois University in Carbondale, where she was born. Her father was a coal miner, "When there were mines." Her mother was a teacher. She was the first Negro cheer leader the university ever had, helped edit the newspaper, and played in the school band. "My classmates would always take me into their confidence, though they had come from towns that were really Southern." A white girl friend's grandmother objected when she wanted to take Barbara home with her during a vacation. "Don't feel guilty," I told her, "because my grandparents are just as prejudiced as yours are." So she said, "You're all right with me." And I said, "You're okay with me, too."

It's about two in the afternoon. The place is in semidarkness, the electricity having been cut off. Fortunately, the battery of my tape-recorder was fully charged.

They said I hadn't paid my bills on time, and when you don't pay your bills on time, you get a black record. So they don't bother to send you a second notice. They send you your bill and then they send you a final notice. This is the first time mine's been cut off. We're in darkness tonight.

If one of the kids wants to read a book. . .

•Source: From *DIVISION STREET: AMERICA*, by Studs Terkel, pp. 171-178. Copyright © 1967 by Studs Terkel. Reprinted by permission of Pantheon Books, a Division of Random House, Inc.

No, they can't read a book. They can't watch TV. They can't listen to the radio. If we didn't have a watch, we couldn't tell what time it was, because we have electric clocks.

In about three hours, it'll be totally dark. What happens?

They'll light a candle. We'll probably eat by candlelight. And just before it gets dark, they'll straighten up their rooms and change their clothes and lay out their school clothes for tomorrow. And this will be it. Then we'll sit in here, I suppose.

. . .I like children and I always thought that mothers needed a lot of understanding. I wanted to be a social worker because I realized people had so many problems that they needed somebody to listen to them, to talk to them, to help them. This was a busy world. People don't have too much time to listen. They have their own problems.

On a bus, it's easy to strike up a conversation. You don't have to say anything. Just look as though you'd listen and people start telling you, oh, something the boss did, or somebody on the job, or what the landlord, say, or complain about prices—they might've just left a sale, you know. People want to talk; they don't have to know you.

But as far as life in public housing is concerned, I don't think it's a very good place to raise children. They can't make noise, they don't have freedom. They can't do like they can in a small town. They can't lay on the grass, they can't climb trees. . .

The phone rings. It was the first of many such calls during the conversation.

That lady was not of this building. She's in the 5200. I'm chairman of this building's council. She called about the course in cashiering. The Joe Louis Milk Company is offering it. If we could get the people. I told them this is no problem. The women here would like to work. People think people on Aid won't work. If you have three or four children and you really care anything about them, you want to be sure when you go to work that they're taken care of. People will accept your money and say they're taking care of your children, but this is not always so. I think I've talked to about thirteen people who say they want to take the course. But one of the questions they ask: Where's the last place you worked and how long ago? If you haven't worked in six or seven years, who's gonna hire you? (B-r-ring!) Excuse me.

"She just got fired from the CHA—The Chicago Housing Authority. She was a Janitress. Quite a thing going on here—among the row people, too. They have a tough supervisor and they don't want the employees to show any insubordination, whether they're right or wrong. There's a great deal of unemployment. Most people who do construction work, they're off quite a bit. It's seasonal. The others who work at the post office or at some stores don't have too much trouble—I don't think."

The people here want to do something. Last year, we had a garden project, seven acres. They raised their own vegetables and flowers. Quite a few of the ladies canned the vegetables. They were so pretty. Some of them had freezers and they had a whole family dinner. I think they enjoyed it because a lot of the people are from the South. They had done farming and this might have carried them back to, you know, memories.

I don't like the Taylor Homes. It's too crowded, there are not enough activities. Who does the child have to talk to at home? Mother's got other kids. If the father's there, he's working or on relief and this he doesn't appreciate. So who wants to be bothered with some kids? So they say: "Mama, what is this?" And: "I don't know, don't bother me. Go play." So they send them out on the porch or downstairs.

The kids have been shoved outside. So they break windows, destroy property, fight—anything to get some attention. That's how you get the gang leaders. Everybody wants recognition. You get it whichever way you can.

We had some teen-age boys around here that were pretty troublesome. Whenever we were going to have something, I'd ask those boys to keep order so that nobody would start any confusion. They were the ones who started the confusion, so by having them look out for the others, I didn't have any problems. They enjoyed this. And they respect you, too. I started out one evening for class and they were roughhousing downstairs and one of them said: "Cool it, here comes Miz Hayes." I hadn't gotten two feet away and they started again. But at least they let me pass. (Laughs.) This is the same group of boys that I would say: I need your help now.

You push them outdoors, out in the galleries: You can't play ball—no roller skating—you're making too much noise by the window. So they go downstairs: You can't sit on the benches—you're not supposed to be around here. Or somebody throws something down: You're not supposed to be hanging around out front. So what are they gonna do? You can push them so far. Then they're gonna push back. But if they had more to do. . .

The turnover's so fast, you know. In a crowded place like this, you . . . oh, so what? You want to stay by yourself. You don't know about the people next door. You don't know what they're like. A lot of people are kinda suspicious. Maybe they resent conditions. A lot of them are on Aid, a lot of wrong things go on. So they stay to themselves and don't have to worry about it getting out.

Often the elevators don't work. I think we have one working today. It might be working at noon, two o'clock it might be out of order. Three o'clock in the morning it might be out of order, you hear bells ringing. Children and adults get stuck in them. Sometimes they get frightened. Kids know how to stick the elevators. They make a game out of it. They find a kid who is afraid and they stick it.

I've walked fifteen flights quite frequently. When I first moved in, I did it for the exercise. I do it now when I have to. It's not too pleasant. You never know who you're gonna find on the stairs. People from off the street. The stairs are open. You might see people sleeping on the stairs, you might see people gambling on the stairs, you might see drinking on the stairs. You never know what your're gonna meet on the stairway. The light might be on, they might be off. If they're off, you're gonna come up those stairs, you know.

Suppose somebody gets sick, needs a doctor, and the elevator doesn't work?

They're just sick then. You have to take them down the stairs. Or, if somebody really gets sick, they'd just die, I suppose. I often wonder what's gonna happen when we have a fire?* I wonder what's gonna happen when they have a fire and the elevators don't work. It'll take 'em longer. We had one in Taylor when a couple of kids did die because they couldn't get an elevator.

They say kids play on 'em and people put 'em out of order; but if this is the case, certainly CHA should do something about it. Put some operators on. They have repairmen and electricians and maintenance men. Wouldn't it be less expensive for the CHA? Of course, it would be more convenient for the resident.

We can't get cooperation from management. Our council has gotten parents to say: We'll operate the elevators during the day, if they put some people on in the afternoon. We did this for two weeks. And they say you can't do anything with the kids. This is not so. After we had been on there for a week, the kids knew when to come down to the first floor instead of going to the third floor to get on. Came to the first floor, lined up and waited. In a matter of fifteen minutes, we had all the school kids home. It was working out.

We asked management to send some people up in the evening. We were gonna help them, too. The kids had gotten used to seeing the parents down there, so they would straighten up, they wouldn't push. They'd get on and wouldn't touch the buttons. They'd tell you what floor. They didn't scream and just make a piano board out of the buttons as they do now. The management didn't object. They just didn't cooperate. They didn't care.

We had a meeting to find a method of keeping the galleries clean. Kids drop food, paper, and so forth. It wasn't swept. The janitors don't do it. They make good money—$441, assistant janitor, it used to be. I think they got a raise. The head janitor makes five something a month. We don't know what for. They don't sweep the galleries, they don't keep the laundry rooms clean. In this building, we have a defective incinerator. When it's smoking, you can't put anything down. You've got two choices: either leave the garbage out there or bring it back into your apartment. Most people leave it out there.

We tried to get the people to keep the porches clean, and we asked management to go along with this little story: Everybody would get fined five or

---

*From the Chicago *Daily News,* December 11, 1965: "Two children were killed Friday in a $200 fire in a Robert Taylor Homes building at 4352 S. State. The victims, Timothy Larde, 5, and his sister, Regina, 3, suffocated in their tenth-floor apartment, firemen said. Neighbors pulled Regina's body from the apartment. Timothy's body was found under a pile of clothes in a closet where he apparently sought refuge. The children's mother was at work, firemen said, and a girl who came to baby-sit discovered the fire. It was confined to the Larde apartment." This item appeared months after our conversation.

ten dollars, everybody on the floor, if there was any debris found outside. We circulated this story through the building for about two or three weeks. You could tell the difference. But there was never this follow-up letter from management. The people would call them to see if this was so. Nothing. They enforce the rules they want to enforce. Just like we say: Don't overcrowd the elevators; nothing was said or done by them until one of the painters was injured and then *they* said: Don't overcrowd the elevators. Until someone from management is affected, there's nothing done.

People would care for things, would keep things clean, if they were given encouragement.

You tell people: Keep your porch clean. You're subject to hear this: "Who are you to tell me what to do? You live here just like I do. Who gave you authority? So they're paying you now?" But management can say: You pay more rent or you get a five-day notice. Can't they say this, too, if it's for the betterment?

If kids come out of this environment stable, emotionally stable, they're lucky. It's very difficult. All the pressures, you know. Say, I live on the fifteenth floor. You can't make too much noise, the people downstairs complain. You can't go out to play. You have to study, watch your little sister. Be careful, don't talk to a stranger. It's an open neighborhood, anybody can come through. The building is like a street. Not just your friends and your bill collectors, but everybody's you know . . . salesmen, anybody.

The people here don't know how to resist. They'll come by with something they know you want, or maybe you need it, you know, but you really can't afford it. And they high-pressure people into taking it. On the installment. No money down, or a dollar down, dollar a week, make it sound so simple. Yeah, they always pay more than what it's worth, you know. Sometimes they can't keep up with the bill and then comes this pressure from the company. They take it away or they pressure you about making a payment. So you have to take your food money or something else to pay them. They want to get paid, they don't really care. You pay for bein' poor.

I don't have time to worry about the H-Bomb. Sometimes you say so throw the Bomb. Who has time to think about the Bomb, when you have to think of how the kids are doing at school, are they gonna make it home all right? Are the elevators working? Where's the next pair of shoes comin' from? You gotta take one of them to the dentist's. Somebody's gotta go to the clinic. These things. Are you gonna . . . how's this kid gonna turn out, you know.

About 3:15. The older children have been coming home from school, one at a time. A burst of energy and news. Suddenly entranced by the presence of the tape-recorder on the kitchen table.

Was it '62 when there was so much talk about the Bomb? We talked about these air-raid shelters and so forth. What are people in Taylor Homes supposed to do? I said: Is this a booby trap? Is this one way of getting rid of a whole lot of Negroes? So you press a button and they're all gone. Too many people in this world, they have to do something about this. This is a good way to get rid of a lot of people.

Time out: Each of the kids has his moment of glory at the microphone. Name, grade, favorite subject, snatches of songs. Much giggling and clowning as voices are played back.

I guess the Bomb bothers a foreigner more than it would me, because the people who have been exposed to war, you know. We see it on TV, it's just a movie. To me, it's just a story. If we get bombed, a lot of people will get killed. Yet, it's a threat. But you just don't worry about it . . . There's nothing I can do about it. Other people control those things anyway.

Who?

The politician. They have a job to do. The people in power tell them what to do.

Who are the people in power?

The people with the money. People that control the hiring and firing, those people. Take Taylor Homes, just one place, for instance. Okay, it's election day: Are you gonna vote? Yeah. No, I'm not gonna be bothered. For what? If agencies would educate the people on their rights, it would be a lot different.

It's not a cause for Negroes. It's a cause for everybody. I don't feel you can keep one person down without keeping yourself down. You can't think of being down and progress at the same time. You got to think one way at a time.

What do you want most in life?

For my children to get a good education—where they'll not have to be pitiful on public aid. And I hope to move from Taylor Homes into a place where I can have my own day-care center. Not just a baby-sitting thing. Expose them to the fine arts. And take them on trips.

Beauty. It's really all around. You just have to find it. Look around and see. Some think the sun and a bright day, that's beauty. Not necessarily. I'm quite sure *this* would be a scene of beauty to an artist. With his canvas. He can see so many things we don't see, you know.

POSTSCRIPT

The elevator going down was crowded. A number of jerky stops. Two women, with Deep South accents, forlorn, lost. Their plaint: the obstinate elevator failed to stop at the fourteenth floor; they had pressed the button again and again; they had been going up and down, three times; yo-yos. "Lotsa time it don't work. I walk up fourteen floors, more'n I can count." One talking, the other nodding: "'At's the truth, Lawd." Young Negro, in the hard hat of a construction worker, gets off at seventh floor. His departing comment over shoulder: "President's Physical Fitness Program." There was no laughter.

The previous selection focused upon the exodus from country to city—or, more correctly, to ghetto. To Paul Goodman, the plight of the newly urbanized Black poor is symptomatic of the fact that all Americans are being impoverished, culturally and psychologically as well as economically, by an "emergency of excessive urbanization." Goodman, author, educator, social and literary critic, argues that this has occurred because our social attitudes and governmental policies treat technology and urbanization as inevitable, inexorable, and beneficial. Goodman, as is his wont, dissents vigorously and offers some interesting proposals. He argues that the great cities can be salvaged and the national life be made more truly livable through the redispersal of population and the reconstruction of rural areas and small towns—thus offering alternative ways of life and counter-vailing systems of political power. In making these suggestions, Goodman raises some fundamental questions concerning the "Technetronic Society." Are urbanization, centralization, and depersonalization inevitable, unchangeable tendencies or can they be reversed by enlightened planning and policy? Are the forces of the Technological Revolution blind, deterministic, and irresistible, or can they be made subject to human control and direction?

Paul Goodman on the City and the Countryside •

1.

I started the last lecture by pointing out how the present style of technology is regarded as an autonomous cause of history. It is even more so with urbanization. It is as if by a law of Nature—the favored metaphor is that the City is a Magnet—that by 1990, 75 percent of the Americans will live in dense metropolitan areas. At present only about 6 percent are listed as rural.

Yet it is not the case that urbanization is a technical necessity. On the contrary, the thrust of modern technology, e.g., electricity, power tools,

•Source: From LIKE A CONQUERED PROVINCE, by Paul Goodman pp. 77–97. Copyright © 1966, 1967 by Paul Goodman. Reprinted by permission of Random House, Inc.

automobiles, distant communication, and automation would seem to be toward disurbanization, dispersal of population and industry. This was the thinking of Marx and Engels, Kropotkin, Patrick Geddes, Frank Lloyd Wright, and other enthusiasts of scientific technology.

Nor is the urbanization a necessity of population growth. In fact, with the bankruptcy of small farming, vast beautiful regions have been depopulating and sometimes returning to swamp. American population growth is supposed to level off, in fifty years, at three hundred million, not a crowded number for such a big area. Yet the cities already show signs of overpopulation. They do not provide adequate city services and probably cannot provide them; they are vulnerable to urban catastrophes that might destroy thousands; it is prohibitively costly to live decently in them; and in my opinion, though this is hard to prove, the crowding is already more than is permissible for mental health and normal growing up.

But it is as with the misuse of technology: the urbanization is mainly due not to natural or social-psychological causes, but to political policy and an economic style careless of social costs and even money costs. Certainly cities are magnets, of excitement and high culture, as markets, centers of administration, and arenas to make careers; but these classical functions of cities of a hundred thousand, capitals of their regions or nations, do not explain our sprawling agglomerations of many millions with no environment at all—Metropolitan New York City has fifteen million and most people cannot get out of it. In general, magnet or no magnet, average people have been content to remain in the provinces and poor people never leave the land, unless they are driven out by some kind of enclosure system that makes it impossible to earn a living. Especially today, when the great American cities are morally and physically less and less attractive, the towns and farms, equipped with TV, cars, and small machines that really pay off in the country, are potentially more and more attractive.

Like the rest of our interlocking system, the American system of enclosure has been an intricate complex. National farm subsidies have favored big plantations which work in various combinations with national chain grocers who now sell 70 percent of food—a hundred companies more than 50 percent. Chains and processors merge. The chains and processors have used the usual tactics to undercut independents and cooperatives. In the cities federally-financed urban renewal has bulldozed out of existence small vegetable stores and grocers who are replaced by the chains. Shopping centers on new subsidized highways bypass villages and neighborhoods. Guaranteed by federal mortgages, real-estate promoters transform farmland into suburbs. Farmer's markets disappear from the cities. As rural regions depopulate, railroads discontinue service, with the approval of the Interstate Commerce Commission. Rural schools are encouraged to degenerate, and land-grant colleges change their curricula toward urban occupations. The Army and Navy recruit apace among displaced farmboys (as they do also among city Negroes and Spanish-speaking Americans).

All this, which sounds like Oliver Goldsmith and Wordsworth, is rationalized by saying, as usual, that it is efficient. One farmer can now feed thirty people. Yet strangely, though most of the farmers are gone and the take of the remaining farmers indeed tends to diminish every year, the price of food is *not* lower, it is about the same. (Let us overlook the inflationary rise due to the Vietnam war.) The difference goes to the processors, packagers, transporters, and middlemen. These operate in the established style. That, is the urbanization and rural depopulation is not technical nor economic but political. The remarkable increase in technical efficiency would just as well produce rural affluence or a cooperative society of farmers and consumers.

It has certainly not been technically efficient to bulldoze the garden land of the missions of Southern California into free-ways, aircraft factories, and smog-choked suburbs, and then to spend billions of public money to irrigate deserts, robbing water from neighboring regions. The destruction of California is probably our most extreme example of bad ecology, but it is all of a piece with the national destruction of the fish and trees, the excessive use of pesticides, the pollution of the streams, the strip mining of the land.

Of course, the galloping urbanization has been worldwide and it is most devastating in the so-called underdeveloped countries that cannot afford such blunders. Here the method of enclosure is more brutal. Typically, my country or some other advanced nation introduces a wildly inflationary standard, e.g., a few jobs at $70 a week when the average cash income of a peon is $70 a year. If only to maintain their self-respect, peasants flock to the city where there are no jobs for them; they settle around it in shantytowns, and die of cholera. They used to be poor but dignified and fed, now they are urbanized, degraded, and dead. Indeed, a striking contrast between the eighteenth-century enclosures and our own is that the dark Satanic mills needed the displaced hands, whereas we do not need unskilled labor.

In the United States, though we collect the refugees in slums, we do not permit them to die of starvation or cholera. But I am again puzzled at the economics of the welfare procedure. Consider the Puerto Ricans. First, for sixty years, by a petty mercantilism worthy of George III, we destroy their agriculture and the possibility of building a broad-based industry; more recently we import a superficial top crust of our own kind of industry, using cheap Puerto Rican labor, and perhaps a third of the people are on relief of some kind; then we allow six hundred thousand Puerto Ricans, a majority with some rural background, to settle in New York City, the most expensive and morally strange possible environment, rather than bribing them to disperse. At present, every week a thousand come to New York and a thousand flee back home to avoid the degradation and narcotics of New York. Similarly, when sharecropping fails in the South, rather than subsidizing subsistence farming and making a try at community development, we give relief money and social work in Chicago and

Los Angeles. Take it at its crudest level: if the cheapest urban public housing costs $20,000 a unit to build, and every city has a housing shortage, would it not be better to give farmers $1000 a year for twenty years, just for rent, to stay home and drink their own water?

2

Partly our urban troubles spring from no planning, partly from just the planning that there is. When concrete observation and sympathy for human convenience are called for, there is no forethought and we drift aimlessly; when there is planning, it is abstract and aimed at keeping things under control or making a profit. For instance, I just contrasted "$20,000 a unit" with "staying home," but no such equation could occur in urban planning, for the word "home" has ceased to exist; the term is "dwelling unit" or "d.u." D.U. is analyzed to meet certain biological and sociological criteria, and it is also restricted by certain rules, e.g., in public housing one cannot nail a picture to the wall, climb a tree in the landscaping, keep pets, engage in immoral sex, or get a raise in salary beyond a certain level. There is a theory, as yet unproved, that planning can dispense with the concept "home"; it is a debased version of Le Corbusier's formula that a house is a machine for living, equal for any tenant and therefore controlled to be interchangeable. But is it the case that people thrive without an own place, unanalyzable because it is the matrix in which other functions occur and is idiosyncratic? Maybe they can, maybe they can't. My point is not that the d.u. is not so good as the shack in a white supremacist county down South; it is better. But that home in a shack plus $1000 a year to improve it is much better even down South, where money talks as loud as elsewhere.

Another term that has vanished from planning vocabulary is "city." Instead there are "urban areas." There is no longer an art of city planning but a science of urbanism, which analyzes and relates the various urban functions, taking into account priorities and allocating available finances. There is no architectonic principle of civic identification or community spirit which the planner shares as a citizen and in terms of which he makes crucial decisions, including necessary uneconomic choices. Such a principle is perhaps unrealistic in a national culture, economy, and technology; we are citizens of the United States, not of New York City. In planning, the interstate and national highway plan will surely be laid down first, and local amenity or existing situations must conform to it; and Washington's ideas about the type of financing and admistration of housing will surely determine what is built. (Oddly, just in such urban functions as highways and housing, local patriotism and neighborhood feeling suddenly assert themselves and exert a veto, though rarely providing a plan of their own.)

But is it the case that urban areas, rather than cities, are governable? Every municipality deplores the lack of civic pride, for instance in littering and vandalism, but it is a premise of its own planning. Anomie is, primarily, giving up on the immediate public environment: the children are bitten by rats, so why bother? the river stinks, so why bother? This kind of depression can go as far as tuberculosis, not to speak of mental disease. In my opinion it is particularly impossible for the young to grow up without a community or local patriotism, for the locality is their only real environment. In any case, when the going gets rough, which happens more and more frequently in American cities, poor people retreat into their neighborhoods and cry, "It's ours!" or they burn them because they are not citizens in their own place. The middle class, as usual, makes a more rational choice: since the center offers neither home nor city nor an acceptable environment for their children, they leave it, avoiding its jurisdiction, taxes, and responsibilities, but staying near enough to exploit its jobs and services.

It is painfully reminiscent of imperial Rome, the return of the farmland to swamp and the flight of the *optimati* from the city center. The central city is occupied by a stinking mob who can hardly be called citizens, and the periphery by the knights and senators who are no longer interested in being citizens. This is an urban area.

3

Moral defects are disastrous in the long run; but American cities are also vulnerable to more immediate dangers, to life and limb. In my own city of New York during the past year we have been visited by ten critical plagues, some of them temporary emergencies that could recur at any time, some abiding sores that are getting worse. It is interesting to list these and notice the responses of the New Yorkers to them.

There was a power failure that for a few hours blacked out everything and brought most activity to a stop. There was a subway and bus strike that for almost a month slowed down everything and disrupted everybody's business. There was a threatened water shortage which persisted for four years and which, if the supply had really failed, would probably have made the city unlivable. In these "objective" emergencies, the New Yorkers responded with fine citizenship, good humor, and mutual aid. By and large, they remember the emergencies as better than business as usual. Everybody was in the same boat.

By contrast, during the long heat wave of last summer—no joke in the asphalt oven of a giant city—there was less enthusiasm. In Chicago it was the occasion for a bad race riot when a fire hydrant was shut off in a Negro neighborhood but (it was said) not in a white neighborhood. In New York it came out that the poor in public housing could not use air conditioners because of inadequate wiring. Evidently, everybody was not in the same boat.

The rivers and bays are polluted and often stink; in a huge city with no open space and few facilities for recreation, this is a calamity. The air is bad; there have just been two alerts within two weeks. The congestion is critical. Traffic often hardly moves, and new highways will only make the situation worse; there is no solution except to ban private cars, but no politician has the nerve to do it. As for human crowding, it is hard to know at what density people can no longer adapt, but there must be a point at which there are too many signals and the circuits become clogged, and at which people do not have enough social space to feel self-possessed. In some areas, in my opinion, we have passed that point. In Harlem, there are sixty-seven thousand to the square mile; people live two and three to a room; and the average child of twelve will not have been half a mile from home.

Toward these abiding ills the attitude of New Yorkers is characteristically confused. They overwhelmingly, and surprisingly, vote a billion dollars to clean up the pollution of the Hudson; they cooperate without grumbling with every gimmick to speed up traffic (though taxi drivers tell me most of them are a lot of nonsense); they are willing to pay bigger bills than any other city for public housing and schools. As people they are decent. But they are entirely lacking in determination to prevent the causes and to solve the conditions; they do not believe that anything will be done, and they accept this state of things. As citizens they are washouts.

Finally, there are the plagues that indicate breakdown, psychopathology, and sociopathology. There are estimated seventy thousand narcotics addicts, with the attendant desperate petty burglary. The juvenile delinquency starts like urban juvenile delinquency of the past, but it persists into addiction or other social withdrawal because there is less neighborhood support and less economic opportunity. Families have now grown up for several generations dependent on relief, reformatories, public hospitals, and asylums as the normal course of life. A psychiatric survey of midtown Manhattan has shown that 75 percent have marked neurotic symptoms and 25 percent need psychiatric treatment, which is of course unavailable.

Given the stress of such actual physical and psychological dangers, we can no longer speak in urban sociology merely of urban loneliness, alienation, mechanization, delinquency, class and racial tensions, and so forth. Anomie is one thing; fearing for one's life and sanity is another. On the present scale urbanization is a unique phenomenon and we must expect new consequences. To put it another way, it becomes increasingly difficult for candid observers to distinguish among populist protest, youth alienation, delinquency, mental disease, civil disobedience, and outright riot. All sometimes seem to be equally political; at other times, all seem to be merely symptomatic.

4

Inevitably, the cities are in financial straits. (At a recent Senate hearing Mayor Lindsay of New York explained that to make the city "livable" would require $50 billion in the next ten years over and above the city's normal revenue.) Since the cities are not ecologically viable, the costs for services, transportation, housing, schooling, welfare, and policing steadily mount with diminishing benefits. Meantime, the blighted central city provides less revenue; the new middle class, as we have seen, pays its taxes in suburban counties; and in the state legislatures the rural counties, which are over-represented because of the drastic shift in polulation, are stingy about paying for specifically urban needs, which are indeed out of line in cost. Radical liberals believe, of course, that all troubles can be immensely helped if urban areas get much more money from national and state governments, and they set store by the reapportionment of the state legislatures as ordered by the Supreme Court. In my opinion, if the money is spent for the usual liberal social engineering, for more freeways, bureaucratic welfare and schooling, bulldozing urban renewal, subsidized suburbanization, and technologized police, it will not only fail to solve the problems but will aggravate them, it will increase the anomie and the crowding.

The basic error is to take the present urbanization for granted, both in style and extent, rather than to rethink it. (1) To alleviate anomie, we must, however "inefficient" and hard to administer it may be, avoid the present massification and social engineering; we must experiment with new forms of democracy, so that the urban areas can become cities again and the people citizens. I shall return to this subject in the following lectures. But (2) to relieve the absolute overcrowding that has already occurred or is imminent, nothing else will do but a certain amount of dispersal, which is unlikely in this generation in the United States. It involves rural reconstruction and the building up of the counrty towns that are their regional capitals. (I do not mean New Towns, Satellite towns, dormitory towns, or other supersuburbs.) In Scheme II of *Communitas,* my brother and I have fancifully sketched such a small regional city, on anarcho-syndicalist principles, as a symbiosis of farm and city activities and values. (Incidentally, Scheme II would make a lot of sense in Canada.) But this is utopian. In this lecture let me rather outline some principles of rural reconstruction for the United States at present, during a period of excessive urbanization.

Liberals, when they think about urbanization, either disregard the country or treat it as an enemy in the legislature. A result of such a policy is to aggravate still another American headache, depressed rural areas. The few quixotic friends of rural reconstruction, on the other hand, like Ralph Borsodi and the people of the Green Revolution, cut loose from urban problems altogether as from a sinking ship. But this is morally unrealistic, since in fact serious people cannot dissociate themselves from the main problems of society; they would regard themselves as deeply useless—just as small farmers do consider themselves. A possible basis of rural reconstruction, however, is for the country to help with urban problems, where it can more cheaply and far more effectively, and thus to become socially important again. (A possible example is how the Israeli *kibbutzim* helped with the influx of the hundreds of thousands of Oriental Jews who came destitute and alien.)

Radicals, what I have called the wave of urban populists, the students, Negroes, radical professors, and just irate citizens, are on this subject no better than the liberals. They are busy and inventive about new forms of urban democracy, but they are sure to call the use of the country and rural reconstruction reactionary. Typically, if I suggest to a Harlem leader that some of the children might do better boarding with a farmer and going to a village school, somewhat like children of the upper middle class, I am told that I am downgrading Negroes by consigning them to the sticks. It is a courious reversal of the narrowness of the agrarian populism of eighty years ago. At that time the farmers lost out by failing to ally themselves with city industrial workers, who were regarded as immoral foreigners and coolie labor. Now farmers are regarded as backward fools, like one's sharecropper father. Although the urban areas are patently unlivable, they have narrowed their inhabitants' experiences so that no other choice seems available.

In Canada a more rational judgment is possible. You have a rural ratio—15 to 20 percent, including independent fishermen, lumberers, etc.—that we ought to envy. Your cities, though in need of improvement, are manageable in size. There is still a nodding acquaintance between city and country. I urge you not to proceed down our primrose path, but to keep the ratio you have and, as your technology and population grow, to work out a better urban-rural symbiosis.

5

Traditionally, in the United States, farming-as-a-way-of-life and the maintenance of a high rural ratio have been regarded as the source of all moral virtue and political independence; but by and large public policy has tended to destroy them. The last important attempt to increase small farming was during the Great Depression when subsistence farms were subsidized as a social stabilizer, preferable to shanties in the park and breadlines. The program lapsed with the war-production prosperity that we still enjoy, and for twenty years, as we have seen, public policy has conspired to liquidate rural life completely. There were no Jeffersonian protests when President Johnson declared three years ago that it was his intention to get two million more families off the land. (More recently, as our surpluses began to run out, LBJ called for a massive return to the land!) Nevertheless, in the present emergency of excessive urbanization, let me offer four or five ideas for rural reconstruction.

(1) At once reassign to the country urban services that can be better performed there, especially to depopulating areas, to preserve what there is. And do this not by setting up new urban-run institutions in the country, but by using local families, facilities, and institutions, adminstered by the now underemployed county agents, Farmers Union, 4-H Clubs, and town governments. Consider a few examples:

For a slum child who has never been half a mile from home, a couple of years boarding with a farm family and attending a country school is what anthropoligists call a culture-shock, opening wide the mind. The cost per child in a New York grade school is $850 a year. Let us divide this sum equally between farmer and local school. Then, the farmer gets $30 a week for three boarders (whom he must merely feed well and not beat), and add on some of the children's welfare money, leaving some for the mothers in the city. With a dozen children, $5000, the underused school can buy a new teacher or splendid new equipment. Add on the school lunch subsidy.

In New York City or Chicago $2500 a year of welfare money buys a family destitution and undernourishment. In beautiful depopulating areas of Vermont, Maine, or upper New York State, or southern Iowa and northern Wisconsim, it is sufficient for a decent life and even owning a house and land. (Indeed, if we had a reasonable world, the same sum would make a family quite well-to-do in parts of Mexico, Greece, or even Ireland.)

The same reasoning applies to the aged. Given the chance, many old people would certainly choose to while away the years in a small village or on a farm, where they would be more part of life and might be useful, instead of in an institution with occupational therapy.

Vacations are an expensive function in which the city uses the country and the country the city. In simpler times, when the rural ratio was high, people exchanged visits with their country cousins or sent the children "to the farm." At present, vacations from the city are largely spent at commercial resorts that tend to destroy the country communities rather than to support them. There are many ways to revive the substance of the older custom, and it is imperative to do so in order to have some social space and escape, which needless to say, the urbanized resorts do not provide.

Here is a more touchy example: the great majority of inmates in our vast public mental institutions are harmless themselves but in danger on the city streets. Many, perhaps most, rot away without treatment. A certain number would be better off—and there would be more remissions—if they roamed remote villages and the countryside as the local eccentrics or loonies, and if they lived in small nursing homes or with farm families paid well to fetch them home. (I understand that this system worked pretty well in Holland.)

(2) Most proposals like these, however, require changes in jurisdiction and administrative purpose. A metropolitan school board will not give up a slum child, though the cost is the same and the classrooms are crowded. No municipality will pay welfare money to a non-resident to spend elsewhere. (I do not know the attitude in this respect of frantically overworked mental hospitals, which do try to get the patients out.) Besides, often in the American Federal system, one cannot cross state lines: a New York child would not get state education aid in Vermont.

So, in conditions of excessive urbanization, let us define a "region" symbiotically rather than economically or technologically. It is the urban area in the surrounding country *with a contrasting way of life and different conditions* that can therefore help solve urban human problems. This classical conception of the capital and its province is the opposite of usual planning. In terms of transportation and business, planners regard the continuous conurbation from north of Boston to Washington as one region and ask for authority to override state and municipal boundaries: for tax purposes, New York would like authority to treat the suburban counties as part of the New York region. These things are, I think, necessary; but their effect must certainly be to increase the monstrous conurbation and make it even more homogeneous. If I regard Vermont, northern New Hampshire, upstate New York, and central Pennsylvania as part of the urban area, however, the purpose of the regional authority is precisely to prevent conurbation and strengthen locality, to make the depopulating areas socially important by their very difference.

(3) The chief use of small farming, at present, cannot be for cash but for its independence, simplicity, and abundance of subsistence; and to make the countryside beautiful. Rural reconstruction must mainly depend on other sources of income, providing urban social services and, as is common, part-time factory work. Nevertheless, we ought carefully to re-examine the economics of agriculture, the real costs and the quality of the product. With some crops, certainly with specialty and gourmet foods, the system of intensive cultivation and hot houses serving farmers' markets in the city and contracting with restaurants and hotels is quite efficient: it omits processing and packaging, cuts down on the cost of transportation, and is indispensable for quality. The development of technology in agriculture has no doubt been as with technology in general, largely determined by economic policy and administration. If there were a premium on small intensive cultivation, as in Holland, technology would develop to make *it* the "most efficient."

In our big cities, suburban development has irrevocably displaced nearby truck gardening. But perhaps in the next surrounding ring, now often devastated, small farming can revive even for cash.

(4) National TV, movies, news services, etc., have offset provincial narrowness and rural idiocy, but they have also had a more serious effect of brainwashing than in, at least, the big cities which have more intellectual resistance. Country culture has quite vanished. Typical are the county papers which now contain absolutely nothing but conventional gossip notes and ads.

Yet every region has seventeen TV channels available, of which only three or four are used by the national networks. (I think the Americans would be wise to have also a public national channel like CBC.) Small broadcasting stations would be cheap to run if local people would provide the programs. That is, there is an available community voice if there were anything to say. The same holds for little theatres and local newspapers. I have suggested elsewhere that such enterprises, and small design offices and laboratories, could provide ideal apprenticeships for bright high school and college youth who are not academic by disposition and who now waste their time and the public money in formal schools. These could be adolescents either from the country or the city. (In New York, it costs up to $1400 a year to keep an adolescent in a blackboard jungle.) Perhaps if communities got used to being participants and creators rather than spectators and consumers of canned information, entertainment, and design, they might recall what they are about.

All such cultural and planning activities, including the sociology of the urban services, ought to be the concern of the land-grant college. At present in the United States we have pathetically perverted this beautiful institution. The land-grant college, for "agriculture and mechanics," was subsidized to provide cultural leadership for its region, just as the academic university was supposed to be international and to teach humanities and humane professions. But now our land-grant and other regional universities have lost their community function and become imitations of the academic schools, usually routine and inadequate, while the academic universities have alarmingly been corrupted to the interests of the nation and the national corporations. Naturally, the more its best young are trained to be personnel of the urban system, the more the country is depleted of brains and spirit.

(5) The fruition of rural reconstruction would consist of two things: a strong cooperative movement and a town-meeting democracy that makes sense, in its own terms, on big regional and national issues. More than a century ago Tocqueville spoke with admiration of how the Americans formed voluntary associations to run society; they were engaged citizens. This would seem to be the natural tendency of independent spirits conscious of themselves as socially

important; they can morally afford to pool their resources for their own purposes. At present it is dismaying to see individual farmers, almost on the margin, each buying expensive machinery to use a few days a year, and all totally unable to cooperate in processing or distribution. They are remarkably skillful men in a dozen crafts and sciences, but they are like children. They feel that they do not count for anything. And unable to cooperate with one another, they cannot compete and they do not count for anything. Correspondingly, their political opinions, which used to be stubbornly sensible though narrow, are frightened and parrot the national rhetoric as if they engaged in dialogue and had no stake of their own.

6

To sum up, in the United States the excessive urbanization certainly cannot be thinned out in this generation and we are certainly in for more trouble. In some urban functions, perhaps, like schooling, housing, and the care of mental disease, thinning out by even a few percent would be useful; and the country could help in this and regain some importance in the big society, which is urban. Nevertheless, the chief advantage of rural reconstruction is for its own sake, as an alternative way of life. It could develop a real countervailing power because it is relatively independent; it is not like the orthodox pluralism of the sociologists that consists of differences that make no difference because the groups depend on one another so tightly that they form a consensus willy-nilly.

The Scandinavian countries are a good model for us. By public policy, over a century and a half. They have maintained a high rural ratio; for a century they have supported a strong cooperative movement; and they have devised a remarkably varied and thoughtful system of education. These things are not unrelated, and they have paid off in the most decent advanced society that there is, with a countervailing mixed economy, a responsible bureaucracy, and vigilant citizens.

Probably the most spectacular figure on the American literary scene during the 1960's and early 1970's was Norman Mailer—existentialist prophet, hipster journalist, and "involved" intellectual. He has been preoccupied with the idea that the American dream has turned into a nightmare, and in the present essay, he argues that Megalopolis plays a leading role in that nightmare. Much like Paul Goodman, Mailer is desirous of saving a sense of community in the city while preserving the vitality of the countryside. However, Mailer sees the hope of salvation in the reapplication of technology itself—in the form of a city of super-skyscrapers. In his interesting speculations about the impact of height and mass upon the psyche of twentieth-century man, Mailer raises some fascinating questions about the esthetic possibilities of advanced technology.

Norman Mailer on Rebuilding the Cities*

What is called for is shelter which is pleasurable, substantial, intricate, intimate, delicate, detailed, foibled, rich in gargoyle, false closet, secret stair, witch's hearth, attic, grandeur, kitsch, a world of buildings as diverse as the need within the eye.

What we get is: commodities swollen in price by false, needless and useless labor. Modern architecture works with a currency which (measured in terms of the skilled and/or useful labor going into a building) is worth half the real value of nineteenth-century money. The mechanical advances in construction hardly begins to make up for the wastes of advertising, public relations, building-union covenants, city grafts, land costs, and the anemia of a dollar diminished by armaments and their taxes. In this context the formulas of modern architecture have triumphed, and her bastards—those new office skyscrapers—proliferate everywhere: one suspects the best reason is that modern architecture offers a pretext to a large real-estate operator to stick up a skyscraper at a fraction of the

*Source: Norman Mailer, *The Idol and the Octopus: Political Writings by Norman Mailer on the Kennedy and Johnson Administrations*, pp. 147-150. Copyright © 1968 by Norman Mailer. Reprinted by permission of the author and the author's agent, Scott Meredith Literary Agency, Inc., 580 Fifth Avenue, New York, N.Y. 10036.

138

money it should cost, so helps him to conceal the criminal fact that we are being given a stricken building, a denuded, aseptic, unfinished work, stripped of ornament, origins, prejudices, not even a peaked roof or spire to engage the heavens.

Look at the depth of the problem in the root of the future:

> ... Fifty years from now ... there will be four hundred million Americans, four-fifths of them in urban areas. In the remainder of this century ... we will have to build homes, highways, and facilities equal to all those built since this country was first settled. In the next forty years we must rebuild the entire urban United States.
>
> —Lyndon Johnson

If we are to spare the countryside, if we are to protect the style of the small town and of the exclusive suburb, keep the organic center of the metropolis and the old neighborhoods, maintain those few remaining streets where the tradition of the nineteenth century and the muse of the eighteenth century still linger on the mood in the summer cool of an evening, if we are to avoid a megalopolis five hundred miles long, a city without shape or exit, a nightmare of ranch houses, highways, suburbs and industrial sludge, if we are to save the dramatic edge of a city—that precise moment when we leave the outskirts and race into the country, the open country—if we are to have a keen sense of concentration and a breath of release, then there is only one solution: the cities must climb, they must not spread, they must build up, not be increments, but by leaps, up and up, up to the heavens.

We must be able to live in houses one hundred stories high, two hundred stories high, far above the height of buildings as we know them now. New cities with great towers must rise in the plain, cities higher than mountains, cities with room for 400,000,000 to live, or that part of 400,000,000 who wish to live high in a landscape of peaks and spires, cliffs and precipices. For the others, for those who wish to live on the ground and with the ground, there will be new room to live—the traditional small town will be able to survive, as will the old neighborhoods in the cities. But first a way must be found to build upward, to triple and triple again the height of all buildings as we know them now.

Picture, if you please, an open space where twenty acrobats stand, each locking hands with two different partners. Conceive then of ten acrobats standing on the shoulders of these twenty, and five upon the ten acrobats, and three more in turn above them, then two, then one. We have a pyramid of figures: six thousand to eight thousand pounds is supported upon a base of twenty pairs of shoes.

It enables one to think of structures more complex, of pyramids of steel which rise to become towers. Imagine a tower half a mile high and stressed to bear a vast load. Think of six or eight such towers and of bridges built between them, even as huge vines tie the branches of one high tree to another; think of groups of apartments built above these bridges (like the shops on the Ponte Vecchio in Florence) and apartments suspended beneath each bridge, and smaller bridges running from one complex of apartments to another, and of apartments suspended from cables, apartments kept in harmonious stress to one another by cables between them.

One can now begin to conceive of a city, or a separate part of a city, which is as high as it is wide, a city which bends ever so subtly in a high wind with the most delicate flexing of its near-to-numberless parts even as the smallest strut in a great bridge reflects the passing of an automobile with some fine-tuned quiver. In the subtlety of its swayings the vertical city might seem to be ready to live itself. It might be agreeable to live there.

The real question, however, has not yet been posed. It is whether a large fraction of the population would find it reasonable to live one hundred or two hundred stories in the air. There is the dread of heights. Would that tiny pit of suicide, planted like the small seed of murder in civilized man, flower prematurely into breakdown, terror and dread? Would it demand too much of a tenant to stare down each morning on a flight of 2,000 feet? Or would it prove a deliverance for some? Would the juvenile delinquent festering in the violence of his monotonous corridors diminish in his desire for brutality if he lived high in the air and found the intensity of his inexpressible and murderous vision matched by the chill intensity of the space through a fall?

That question returns us to the perspective of twentieth-century man. Caught between our desires to cling to the earth and to explore the stars, it is not impossible that a new life lived half a mile in the air, with streets in the clouds and chasms beyond each railing could prove nonetheless more intimate and more personal to us than the present congestions of the housing-project city. For that future man would be returned some individuality from his habitation. His apartment in the sky would be not so very different in its internal details from the apartments of his neighbors, no more than one apartment is varied from another in Washington Square Village. But his situation would now be different from any other. His windows would look out on a view of massive constructions and airy bridges, of huge vaults and fine intricacies. The complexity of our culture could be captured again by the imagination of the architect: our buildings could begin to look a little less like armored tanks and more like clipper ships. Would we also then feel the dignity of sailors on a four-master at sea? Living so high, thrust into space, might we be returned to that mixture of awe and elation, of dignity and self-respect and a hint of dread, that sense of zest which a man must have known working his way out along a yardarm in a stiff breeze at sea? Would the fatal monotony of mass culture dissolve a hint before the quiet swaying of a great and vertical city?

Come mothers and fathers,
Throughout the land
And don't criticise
What you can't understand
Your sons and your daughters
Are beyond your command
Your old road is rapidly agin'
Please get out of the new one
If you can't lend your hand
For the times they are a-changin'.

Bob Dylan●

IV. Technology's Children

To the staid members of Middle America, one of the most shocking phenomena of the turbulent 1960's and 1970's was the abrupt rise of the Youth Movement. Youth, creating its own life styles, hair styles, music styles, and protest styles, suddenly seemed light-years removed from the generally apathetic, career-oriented "cleancut kids" of the placid 1950's. At roughly the same time, the universities lost their ivory-tower calm, as riots, burnings, bombings and massive disruptions began to plague them. College students, heretofore considered a coddled semi-aristocracy, embraced a bewildering sequence of causes—civil rights, anti-war, Black Power, Women's Lib, ecology—with a militant passion that seemed about to upset the liberal democratic consensus as well as the social order. Some observers announced with glee that all of this indicated the onset of "The Revolution." Others replied with disgust that it signified "the tyranny of spoiled brats" resulting from permissive child-rearing.

●Source: THE TIMES THEY ARE A-CHANGIN' © 1963 by M. WITMARK & SONS. Used by permission of WARNER BROS. MUSIC. All Rights Reserved.

To others, such as Kenneth Keniston, a psychologist and student of contemporary youth, "the Movement" must be understood in terms of the impact of advanced technology and the resultant acceleration of social change. Just as the abundance created by the industrial revolution made it possible for the young to enjoy the stage of adolescence, argues Professor Keniston, so in our era, ". . . a post-adolescent stage of youth is beginning to be made available by post-industrial society." Mainly because of the Technological Revolution, the pace and scope of social, political, economic, and intellectual change in the twentieth-century has been unprecedented in human history—and has, in fact, created a sense of historical dislocation. The result, for young Americans, has been a paradoxical condition of affluence, tension, and alienation. With these phenomena in mind, and drawing upon his studies of youthful radicalism in the "Vietnam Summer" project of 1967, Kenneth Keniston analyzes "the Movement" in the following essay.

Kenneth Keniston on Change, Affluence, and Violence[*]

The young radicals I interviewed were born near the end of the Second World War, and their earliest memories date from the years just after it. Their parents were born around the time of the First World War; their grandparents are, without exception, the children of the nineteenth century. Their parents are thus members of the first modern generation to emerge from the Victorian era. And these young radicals are the first products of the post-war world, the first post-modern generation. In tracing the story of their lives, I have discussed the personal meaning of three central themes: change, affluence, and violence; in each, the psychological, the social, the political, and the historical are fused. And each of these issues was so much a part of the young radicals' lives that it is only by stepping aside to consider the historical ground on which they grew that we can perceive the impact on these lives of the history of the post-war era.

●Source: From *Young Radicals: Notes on Committed Youth,* copyright © 1968 by Kenneth Keniston. Reprinted by permission of Harcourt Brace Jovanovich, Inc., pp. 229-256.

In the last chapter, I argued that the issue of change is pervasive in the development of these young men and women. Despite their underlying ties to their personal and familial pasts, their development has involved major alterations, reversals, and reassimilations of that past. As young adults, they remain acutely aware of how far they have come, of the differences between their generation and their parents'. More than that, they have in their own lives witnessed and experienced social and historical changes on an unprecedented scale, lived through the Cold War, the McCarthy era, the Eisenhower period, the short administration of Kennedy and the long one of Johnson. By becoming involved with the New Left, they have linked themselves to a moving, changing movement of dissenting youth. And as individuals, even in their early adulthood, they remain open to the future, eager to change, "in motion."

Similarly, the fact of affluence is crucial to their lives. Not one of these young men and women comes from a background of deprivation, poverty, discrimination, or want. From their earliest years they have simply taken for granted that there would be enough—not only enough to survive, but enough for a vacation every year, a television set, a family car, and a good education. They grew up in a world where they and virtually everyone they knew took prosperity and the luxuries it provides most Americans totally for granted. Until they reached adolescence and social consciousness, few of them were immediately aware of the facts of poverty, discrimination, and hunger. Their affluence provided them not only with economic security, but with the preconditions for the independence they exhibit in later life: families generally free from acute anxiety over status, thoughtful and well-educated parents, schools and colleges that—whatever their limitations—exposed them to many of the riches of world tradition, and the extraordinary privilege of a lengthy adolescence and youth in which to grow, to become more complex, to arrive at a more separate selfhood.

The issue of violence, and of the fear and anger it inspired, starts with the earliest memories of many of these young radicals. Recall the young man whose first memory involves his backyard parade at the end of World War II, and whose second memory is of his hysterical terror at the encyclopedia pictures of an atomic-bomb explosion and an army tank. Remember the angry and menacing mob in one early memory, the jealous rage at a younger brother in another, the "gruesome" fights in the playground in still another. Such early memories, of course, mean many things. They point to themes of lifelong importance; they can serve as a "screen" for other less conscious issues—as symbolic alternatives to what is not remembered—and they indicate something about the fears of the dreamer both when he was small and as an adult. Taken with the rest of what we now know about these young radicals, these memories indicate a special sensitivity to the issue of violence—inner and outer—that continues as a central theme in their lives.

These young radicals, then, are members of the first post-modern generation, and their lives are permeated with the history of the past two decades. They, and I as their interviewer, took such changes completely for granted, and rarely felt compelled to note their occurrence and significance. Indeed, in the last third of the twentieth century, we all take for granted the revolutionarily changing world in which we have lived from birth. Yet to understand better what these radicals have done and are attempting to do, to comprehend the style they are creating, requires that we also examine the historical ground of their development.

CHANGE AND THE CREDIBILITY GAP

The twentieth century, as a whole, has been a period of un-precedentedly rapid social, industrial, ideological, and political change. But during the post-war era, the pace of change has increased still further, transforming the world in a way that no one, twenty-five years ago, could have anticipated. These post-war years have brought to the more advanced nations of the world a kind of affluence rarely even dreamed of before. They have seen the often violent liberation of the majority of the world's population from colonial rule. They have seen a time of extraordinary scientific and technological innovation that has profoundly transformed our physical, human, social, and cultural environment. And no one can foresee the end of change.

In the last two decades, it has become increasingly obvious that extremely rapid social change is endemic to the modern world. It is unnecessary to chronicle in detail the specific changes that have occurred. Suffice it to note that the material and technological changes that are easiest to pinpoint and discuss constitute but a small part of the over-all process of social change. Even more important have been the less tangible, more gradual, often unnoticed yet radical transformations in social institutions, in the ways men relate to each other and their society, in interpretations of the world and of history, and in the definitions of the goals of life itself. Increasingly, we take such changes for granted, welcoming them, accommodating ourselves to them as best we can, growing used to a world where nothing is permanent. Partly for this reason, we have barely begun to understand the human effects of rapid, continual social change. Especially for the post-war generation, who have always known a world of flux and transformation, change is so much a part of life that they seldom reflect on its meaning. It is like the grammar of our language, or the quality of the air, or the face of a family member: we seldom stand back to notice.

Yet the forces that affect us most profoundly are often those we never stop to notice. In *The Uncommitted* I have discussed at greater length some of the human effect of chronic social change. All of these effects are evident in the lives of the young radicals who led Vietnam Summer. Even in these young men and women, for example, we see a gap between the generations, such that each generation must reconsider and re-examine the values of its heritage for itself. The parents of these particular young radicals have been able to establish a continuity in what I have called core values between themselves and their children. In this respect, there is probably *less* of a generational gap in the families of these young radicals than in the families of most of their contemporaries. But this continuity is at the level of basic personal values like honesty and responsibility, rather than at the level of specific political programs and social creeds. Even the children of old radicals simply take it for granted that their political values and goals will be different from those of their parents. As far as formal values are concerned, then, the prime symptom of the generational gap is apparent: both generations take more or less for granted that the public philosophies of parents are largely irrelevant to their children. In a time of rapid value change, it may be that the only possible value continuities between the generations must involve core values so broad, general, and basic that they can remain relevant despite a radically transformed human and social world.

Another corollary of rapid social change is a focus on the present as contrasted with the past and future. As the pace in social change accelerates, the relevance of the past (and of those like parents who are a part of it) decreases; similarly, the predictability and stability of the future as an object of planning lessens. No traditional verity can be accepted without testing its continuing validity: the skills, styles, rules, and truths of the past become quickly old-fashioned. Since the rate of social change is continually accelerating, and since, in the past, most efforts to predict the future have been dismal failures, the possiblity of making concrete plans for the future decreases steadily. Whatever its many other meanings, the focus on the short range and the tactical in the New Left reflects the consciousness of many of today's youths that long-range planning is virtually impossible, given the many imponderables that make the best laid plans go astray. And the absence of utopian visions of the future among young radicals may not reflect a failure of imagination as much as an awareness that the future is simply impossible to anticipate.

Another consequence of a rapidly changing world is the emphasis placed on such psychological qualities as flexibility, openness, adaptability, and personal change. Men always identify themselves with what they take to be the nature of the historical process in which they are immersed: in a time of rapid social and historical change, psychological changeability is therefore stressed. But flexibility is also a way of coping with the demands of the modern historical process. In a stable society, changing individuals must at each stage of their psychological development accommodate themselves to the same static society. But in a rapidly changing society, individuals must adapt themselves at each stage of their personal development to a constantly different physical, human, and social environment. Fixed positions—be they fixed character traits, rigid defenses, absolutely held dogmas, or tenaciously acquired skills—are a commitment to obsolescence. To "keep up with the times," men and women must be ready to change—often radically—throughout their lives. This readiness is, of course, a salient quality in young radicals.

Even the ambivalences of these young radicals toward their parents of the same sex, and the extreme selectivity of their identifications with these parents, are connected to the fact of social change. In an era when the life-situations of children differ so drastically from the environments of their parents as children, simple and "total" identification between generations is rarely possible. Children recognize intuitively that their parents are the products of a different social and historical matrix, and become more selective about following in their footsteps. Parents, in turn, also tend to acknowledge these generational differences, and no longer dare demand the same filial loyalty, obedience, or imitation. Children must learn to winnow the historical chaff from the grain in identifying with their parents, just as these young radicals chose a few core values as their inheritance, rejecting the rest. The particular content of parental identifications among young radicals has many special features, but the need to be selective in identifying is inherent in an era of rapid change.

The major transformations of the past decades also contribute to a widespread sensitivity of today's youth to the *discrepancy between principle and practice,* and may help explain why the charges of insincerity, manipulation, and dishonesty are today so often leveled by the young against the old. During a time when values change with each generation, the values most deeply embedded in parents and expressed in their behavior in times of crisis are often very different from the more "modern" principles, ideals and values that parents profess and attempt to practice in bringing up their children. Filial perception of this discrepancy between parental practice and principle may help explain the very widespread sensitivity amongst contemporary youth to the "hypocrisy" of the previous generation. Among the young radicals interviewed, the schism in the parental image seems related not only to the idiosyncratic behavior of specific parents, but to this broader problem of transmission of values in a time of rapid change.

The grandparents of today's twenty-year-olds were generally brought up during the pre-World War I years, heirs of a Victorian tradition as yet unaffected by the value revolutions of the twentieth century. They reared their own children, the parents of today's youth, in families that emphasized respect, the control of impulse, obedience to authority, and the traditional "inner-directed" values of hard work, deferred gratification and self-restraint. Their children, born around the time of the First World War, were thus raised in families that remained largely Victorian in outlook.

During their lifetimes, however, these parents (and in particular the most intelligent, well educated, and privileged of them) were exposed to a great variety of new values that often changed their formal convictions. During their youths in the 1920's and 1930's, major changes in American behavior and American values took place. For example, the "emancipation of women" in the 1920's, marked by the achievement of suffrage for women, coincided with the last major change in actual sexual behavior in America; during this period, women started to become the equal partners of men, who no longer sought premarital sexual gratification solely with women of a lower class. More important, the 1920's and 1930's were an era when older Victorian values were challenged, attacked, and all but discredited, especially in educated middle-class families. Young men and women who went to college during this period were influenced by "progressive," "liberal," and even psychoanalytic ideas that contrasted sharply with the values of their childhood families. Moreover, during the 1930's, many of the parents of today's upper-middle-class youth were exposed to, or involved with, the ideals of the New Deal, and sometimes to more radical interpretations of man, society, and history. And in the 1940's and 1950's, when it came time to raise their own children, the parents to today's youth were strongly influenced by "permissive" views of child-rearing that again clashed with the techniques by which they themselves had been raised. Thus, many middle-class parents moved during their lifetimes from the Victorian ethos in which they had been brought up to the less moralistic, more humanitarian, and more "expressive" values of their own adulthoods.

But major changes in values, when they occur in adult life, are likely to be less than complete. To have grown up in a family where unquestioning obedience to parents was expected, but to rear one's own children in an atmosphere of "democratic" permissiveness and self-actualization—and never to revert to the practices of one's own childhood—requires a change of values more comprehensive than most adults can achieve. Furthermore, behavior that springs from values acquired in adulthood often appears somewhat forced, artificial, or insincere to the sensitive observer. Children, always the most perceptive observers of their own parents, are likely to sense a discrepancy between their parents' avowed and consciously held values and their "basic instincts," especially with regard to child-rearing. In addition, the parental tendency to "revert to form" is greatest in times of family crisis, which, of course, have the weightiest effect upon children. No matter how "genuinely" parents hold their "new" values, many of them, when the chips are down, fall back on the lessons of their own childhoods.

In a time of rapid social change, then, a *credibility gap* is likely to open between the generations. Children are likely to perceive a discrepancy between what the parents avow as their values and the actual assumptions from which parental behavior springs in times of crisis. In the young radicals interviewed, the focal issue of adolescent rebellion against parents seems to have been just this discrepancy: the children argued that their parents' endorsement of independence and self-determination for their children was "hypocritical" because it did not correspond with the parents' actual behavior when their children seized the independence offered them. Similar perceptions of "hypocrisy" occurred for others around racial matters: there were a few parents who supported racial and religious equality in principle, but became upset when their children dated someone of another race or religion. Around political activity similar issues arose, especially during the 1950's. For example, many of the parents of today's youth espoused in principle the cause of political freedom; but most were not involved in politics themselves and some opposed their children's involvement lest they "jeopardize their records."

Of course, in no society do parents (or anyone else) ever fully live up to their own professed ideals. In every society, there is a gap between creedal values and actual practices; and everywhere the recognition of this gap constitutes a powerful motor for social change. But in most societies, especially when social change is slow and social institutions are powerful and unchanged, there occurs what can be called the *institutionalization of hypocrisy*. Children and adolescents routinely learn when it is "reasonable" to expect that the values parents profess will be implemented in their behavior, and when it is not reasonable. There develops an elaborate system of exegesis and commentary upon the society's creedal values, excluding certain people or situations from the full weight of these values or "demonstrating" that apparent inconsistencies are not really inconsistencies at all. Thus, in almost all societies, a "sincere" man who "honestly" believes one set of values is frequently allowed to ignore them completely, for example, in the practice of his business, in his interpersonal relationships, in dealings with foreigners, in relationships of his children, and so on—all because these situations have been defined by social consensus as exempt from the application of his creedal values.

In a time of rapid social change and value change, however, the institutionalization of hypocrisy tends to break down. "New" values have been in existence for so brief a period that the exemptions to them have not yet been defined, the situations to be excluded have not yet been determined. The universal gap between principle and practice appears without disguise. Thus, the mere fact of a discrepancy between creedal values and practice is not at all unusual. But what is special about the present situation of rapid value change is, first, that parents themselves tend to have two conflicting sets of values, one related to the experience of their early childhood, the other to the ideologies and principles acquired in adulthood; and, second, that no stable institutions or rules for defining hypocrisy out of existence have yet been fully evolved. In such a situation, the young see the Emperor in all his nakedness, recognizing the value conflict within their parents and perceiving clearly the "hypocritical" gap between ideal and behavior.

This argument suggests that the post-modern youth may not be confronted with a gap between parental preaching and practice that is "objectively" any greater than that facing most generations. But they do confront an unusual internal ambivalence within the parental generation over the very values that parents successfully inculcated in their children, and they are "deprived" of a system of social interpretation that rationalizes the descrepancy between creed and deed. It seems likely, then, that today's youth may simply be able to perceive the universal gulf between principle and practice more clearly than previous generations have done.

This points to one of the central characteristics of today's youth in general and young radicals in particular: they insist on taking seriously a great variety of political, personal, and social principles that "no one in his right mind" ever before thought of attempting to extend to such situations as dealings with strangers, relations between the races, or international politics. For example, peaceable openness has long been a creedal virtue in our society, but it has rarely been extended to foreigners, particularly those with dark skins. Similarly, equality has long been preached, but the "American dilemma" has been resolved by a series of institutionalized hypocrisies that exempted Negroes from the application of this principle. Love has always been a formal value in Christian societies, but really to love one's enemies—to be generous to policemen, customers, criminals, servants, or foreigners—has been considered folly.

The fact of social change, then, is not only distantly perceived by those who are growing up, but immediately interwoven with the texture of their daily lives as they develop. Many of the seemingly "special" characteristics of this small group of young radicals are connected not only to the vicissitudes of their individual histories, but to the history of their generation and of the modern world. The tenacity with which these young men and women adhere to a small number of the core values from their early family lives, their short-range plans, their absence of political program and visions of the future, and their enormous emphasis on openness, change, and process is both a reflection of, and a response to, a world changing at a dizzying rate in a direction that no one can foresee.

And these speculations on the credibility gap and the "deinstitutionalization of hypocrisy" in a time of rapid change may help explain two further facts about young radicals: first, they frequently come from highly principled families with whose core principles they continue to agree, but they often see their parents as somehow ineffectual in the practice of these principles; second, they have the outrageous temerity to insist that individuals and societies live by the values they preach. And these speculations may also explain the frequent feeling of many who have worked intensively with today's dissenting youth that, apart from the "impracticality" of some of their views, these sometimes seem to be the only clear-eyed and sane people in a society and a world where most of us are systematically blind to the traditional gap between personal principle and practice, national creed and policy.

THE ADVENT OF AUTOMATIC AFFLUENCE

To any American who has grown up since the Second World War, one of the most important facts of life has been the continually increasing affluence around him. For all middle- and upper-class young Americans, as for increasing numbers of working-class youth, the fact of affluence is simply taken for granted—prosperity has become automatic. For example, although one or two of the young radicals who led Vietnam Summer came from lower-middle-class families and considered themselves "poor" during childhood, questions of income, security, social status, upward mobility, and finding a job were largely irrelevant when the time came for them to consider adult commitments. And when they realized during their adolescences that the affluence they took for granted did not extend to all Americans—much less to the impoverished two-thirds of the world—they reacted with surprise, shock, and dismay. Material prosperity alone has made a difference in the development of this generation. The "luxuries" of an affluent age—electronic communications, rapid transport, good housing, physical comfort, readily available music, art, and literature, good health care and longevity—have helped give this generation its distinctive style. Without material affluence, the restlessness, mobility, and "wastefulness" of today's youth could hardly be understood.

But the impact of affluence extends considerably beyond its material benefits. "Affluence" can stand as shorthand for a variety of other changes in American institutions, the economy, family life, education, and the definition of the stages of life, all of which have affected the outlook of this generation. Material affluence is made possible by a system of production, innovation, and organization that defines the options open to today's young men and women, just as it has been the framework for their development to date. Affluence, in a broad sense, has both opened new doors and closed old ones.

Social criticism in the past decades has emphasized the destructive aspects of technology, bureaucracy, specialization, centralization, and bigness. Yet we have also begun to realize that these ambivalently viewed features of our society may be necessary conditions for the advantages of affluence. Our prosperity is built upon high technology, as upon complex and bureaucratic social organization. And both technology and differentiated social roles involve specialization and technical competence far beyond the basic requisites of literacy and fluency with numbers. Furthermore, in any highly specialized society, complex systems of coordination, social control, and communication must be developed to harmonize the work of specialized role-holders. Even sheer size sometimes increases affluence: centralization not only can permit industrial efficiencies, but sometimes facilitates administrative coordination. The advent of electronic communications and rapid transportation had made it increasingly possible for a small number of men to coordinate and control the activities of vast numbers of their fellows. For better and for worse, then, our affluent society is technological, specialized, bureaucratized, and complexly controlled. In such a society, most educated adults not only do highly specialized work, but are involved in complex networks of social coordination that they must accept if the System is to function smoothly.

All of these characteristics of modern society contribute to the malaise and reluctance of many of today's youth when they confront the System. Yet these same young men and women, like all of us, consider the many benefits of affluence as "givens" of modern life. They take for granted that just as the machine and factory production made possible the industrial revolution by multiplying each man's physical efforts a dozenfold, so now, in the technological era, the computer is increasingly freeing men from routine and repetitive mental work. Men and women need no longer work in the fields or factories from dawn to dusk to produce the requisites for survival. For affluent Americans (who are the majority), survival, subsistence, and starvation are no longer an issue. A small part of the population can produce the essentials of life, while the rest produce goods and services that, to previous generations, would have appeared unprecedented luxuries.

These "luxuries" include not only the material commodities that fill American life, but less tangible opportunities for education, the cultivation of the mind, and the fulfillment of psychological needs beyond the need for subsistence, security, and status. By vastly extending the power and reach of each individual, the affluent society both permits and requires men to be "unproductive" for many years of their lives. The labor of children, adolescents, and, increasingly, post-adolescents is no longer needed by the economy. On the contrary, keeping young men and women off the labor market is a net social gain because it allows fuller employment of their elders. In addition, an affluent society increasingly requires the young to stay off the labor market in order to learn the high technological skills required to maintain affluence. The result, of course, is the historically unprecedented situation of prolonged higher education, extending well into the twenties, for a larger and larger proportion of the American population.

The postponement of entry into the labor force has contributed to a redefinition of the life cycle, underlining the connection between social opportunity and developmental stage. Giving large numbers of young men and women the opportunity to have an adolescence is an achievement of industrial societies. In many preindustrial societies, even childhood was forcibly aborted by the requirement that children begin to work before puberty. When this happens, the full psychological experience of childhood as we define it in modern society is inevitably cut short: children are small adults—by our modern standards, old before their time. But even in those societies where psychological childhood continues until biological puberty, adolescence as a psychological experience is rarely permitted.

To be sure, the physiological changes that announce the possibility of an adolescent experience occur in every society, regardless of what the society chooses to make of these changes. But in most previous societies, only the extraordinarily wealthy, talented, or fortunate were allowed anything like an adolescence. Even the wellborn Romeo and Juliet were thirteen years old; in the Middle Ages, kings assumed their thrones in their teens; and most children of the common people began working in the fields (in later times, in factories) well before they reached puberty. Allowing the possibility of adolescent development is only one possible reaction to the approach of biological adulthood: historically it is a relatively rare reaction. Even today, in primitive societies, puberty rites more often serve to hasten the child toward adulthood than to permit him anything like the possibility of adolescent turmoil, emotional growth, and independence. Although from the beginnings of history, the old have deplored the irreverence of the young, adolescence as a distinctive stage of life that *should* be made available to all young men and women has only begun to be recognized during the past two centuries in advanced societies.

By creating a vast surplus of wealth, modern societies have freed first children and then teen-agers from the requirements of farm and factory labor. Even before the industrial revolution, of course, a small number of young men and women were allowed a deferment of full involvement in adult work. And a few of them—a few from among the pages and princes, novices and apprentices—were sometimes able to create for themselves what we now recognize as an adolescence. But most, lacking opportunity and social sanction, moved directly from childhood to adulthood. The industrial revolution, however, created a new bourgeoisie with a commitment to education as a pathway to success for their children. The new middle class also had the means to allow children freedom from labor after puberty. There began to develop—for the middle classes at least—a vague concept, at first, of a post-childhood, preadult stage of life, a stage of continuing education that was initially modeled after the apprenticeship. Little by little, however, it became clear that this stage of life had qualities of its own. The adolescent gradually emerged as something more than a cross between a child and an adult.

First for the upper middle class, then for the lower middle class, and then, increasingly, for the working-class youth, adolescence became routinely available. And although the precise definition of the expected qualities of the adolescent is sensitive to the particular values of each society, in most affluent societies today, adolescence is recognized as *sui generis,* as important for the fullest possible unfolding of human potentials, and as a right to be guaranteed through compulsory education and anti-child-labor laws.

We should not forget how recently all of this has taken place, nor how incomplete it still is. Some of Marx's most vehement strictures in the middle of the nineteenth century were directed against the use of children in factories. And in America, the child-labor laws were passed only in the twentieth century. For many young Americans, and for an even greater proportion of the young in other nations, the psychological experience of adolescence is still aborted by the failure of education or the assumption of major economic responsibilities in the mid-teens—years that by our modern reckoning are only the beginning of adolescence. For large numbers of the poor, the deprived, the undermotivated, the psychologically or intellectually handicapped, adolescence still does not take place.

Even if it has not yet been extended to all, making the experience of adolescence available to most young men and women in modern society must be counted among the achievements of affluence. The possibility of adolescence as a psychological experience is dependent on economic conditions that free adolescents from the need to work, as upon the development of new values that make child or adolescent labor seem "outrageous" to right-thinking men and women. Only when a society produces enough to liberate young men and women between the ages of twelve and eighteen from labor can it demand that they continue their educations and allow them to continue their psychological development; only then can abhorrence of the "exploitation" of adolescents develop.

Affluence has also permitted changes in the quality of family life, especially among better-educated Americans. During the twentieth century, growing numbers of men and women, responding to the opportunities and demands of industrial society, have at last begun college, with many completing it and continuing on for their Ph.D. Higher education changes the outlooks and styles of at least some of those who pass through it. Its impact is difficult to describe precisely, but at best it allows greater freedom to express underlying feelings and impulses, greater independence of outlook and thought, and increased sympathy for the underdog. Also, since the best educated are generally those who attain greatest affluence in their own lives, higher education indirectly gives its graduates an adult life that is more secure, freer from the struggle for subsistence and status, and more open to the pursuit of non-material self-expressive goals. Educated parents who have attained professional and economic security are in turn able to develop a distinctive family style that has important effects upon children.

Although they themselves may have had to struggle out of poverty, today's well-educated and affluent parents have generally "arrived" by the time they raise their own children. Compared to their own parents, they are more likely to instill in their children the special values of self-actualization—independence, sensitivity to feelings, concern for others, free expression of emotion, openness, and spontaneity. And since such parents tend to have relatively few children, they are able to lavish on each child an enormous amount of individual attention. Upper-middle-class educated women need not work to support the family: most devote themselves entirely to bringing up their small children. Even those who do work are likely to feel restored by their work rather than depleted. All of this means that affluent mothers are increasingly free to devote themselves to their small brood of children. Such devotion can have the bad consequences we see in the familiar stereotype of "Momism." But its good consequences are equally important: in many affluent families, children grow up unusually well cared for emotionally and psychologically, the objects of thoughtful attention and informed devotion. Increasingly, affluent middle-class parents *educate* their children, rather than merely training them. And in some affluent families, one finds a parental devotion to the autonomy, self-determination, and dignity of children that is without precedent, even in American history.

Obviously, not all affluent middle-class families fit this rosy description: such families are clearly in a minority. A full account of the impact of affluence and education of the American family would have to discuss other parental responses, among them family styles that lead to filial apathy, alienation, neurosis, or conformity. But affluence means that families like those I am describing—devoted, principled, expressive, thoughtful, humanitarian, and permissive—are increasing in number. Whatever the other satisfactions they derive from their children, parents in these families genuinely desire them to be independent, thoughtful, honorable, and resourceful men and women. To be sure, in these as all families, parents are full of foibles, contradictions, inconsistencies, and faults. And as I have suggested, in a time of rapid value change, the values that parents attempt to apply in bringing up their children may contrast with the more "instinctive" values that have their roots in the parents' own upbringing.

Yet for all their characteristic faults, the families of the educated and affluent have freed a growing number of today's youth to concern themselves with the welfare of others and the wider society. Their security makes possible an identification with others who are insecure; their affluence permits them to worry about those who are poor; their freedom allows them to care about those who are enslaved. Families like the families of the radicals who led Vietnam Summer are impressively *good*. They have given their children great strength, integrity, and warmth. The devotion to family core values that we see in many young radicals derives from parents who have principles and care lovingly for their children. Even the ability of young radicals to be different from their parents may stem partly from their parents' genuine willingness to let them be different. These are children, then, who have been taught from an early age to value independence, to think for themselves, to seek rational solutions, and to believe that principles should be practiced. As Richard Flacks, one of the most astute observers of the contemporary New Left, has put it, these young men and women are members of a "liberated generation."

This argument suggests that in an affluent society, the psychological and social underpinnings of radicalism have begun to change. In non-affluent societies, radicals and revolutionaries—who almost invariably come from relatively privileged backgrounds—tend to react with guilt to the "discovery" of poverty, tyranny, and misery. Furthermore, many radical and revolutionary groups have in the past sought social and political changes that would improve their own position, giving them freedom, power, or benefits they did not possess. In a society like our own—where affluence, economic opportunity, and considerable political freedom are the rule—radicalism is less likely to be built upon personal feelings of deprivation or a desire to improve one's own position. Nor is the guilt of the wealthy when confronted with the poor as likely a motivation for the radical's commitments. While radical leaders of all eras have typically been men of high principle, the role of principle increases further in an affluent era. The radical's basic goal is not to achieve new freedoms, opportunities, or benefits for himself, but rather to extend to all the freedoms, opportunities, and benefits he himself has always experienced. In an affluent world, the radical feels indignation rather than guilt; outrage rather than oppression.

VIOLENCE: SADISM AND CATACLYSM

The focal issue in Vietnam Summer was ending American involvement in violence in Southeast Asia. And the issue of violence is central not only for young radicals, but for the modern world. Hanging over the lives of all men and women during the past decade has been the Bomb, and the terrifying possibilities of *technological death* it summarizes and symbolizes. These include not only holocaustal destruction by thermonuclear blast and radiation, but the equally gruesome possibilities of the deliberate spread of virulent man-perfected disease or the use of lethal chemicals to destroy the functioning of the human body.

Premature death has, of course, always been a fearful possibility in human life. But technological death is new in a variety of ways. It is now realistic to imagine not only one's own unannounced death and perhaps the death of one's intimates through natural catastrophe, but to envision the "deliberate" destruction of all civilization, all human life, or, indeed, all living things on earth. Furthermore, technological death has a peculiar quality of impersonality, automaticity, and absurdity to it. Until the relatively recent past, most man-inflicted deaths have at least been personal acts: the jealous husband murders his wife's lover, the soldier shoots the enemy on the battlefield, the cannibal kills the member of a neighboring tribe, the sadist butchers his victim. Technological death, in contrast, requires no contact between man and man. One well-intentioned bureaucrat (who means no harm, is only following orders and is doing his duty for his country) can press a button and set in motion a chain of events that could mean the burning, maiming, and death of most of those now alive. Paradoxically, malice, anger, and hostility are no longer necessary to create a cataclysm beyond the imaginings of the darkest sadist. It only takes an understandable inability to visualize the human meaning of a "megadeath."

The technology of death has hung like a sword over the lives of this post-modern generation. Recall, once again, how in the early memories of these young radicals, the violence of the outside world found echo and counterpart in the violence of inner feelings: on the one hand, the atomic bomb, the menacing mob, the gruesome playground fights; on the other hand, rage, fear, and anger. The word "violence" itself suggests both of these possibilites: the *psychological* violence of sadism, exploitation, and aggression, and the *historical* violence of war, cataclysm, and holocaust. In the lives of these young radicals, as in much of their generation, the threats of inner and outer violence are fused, each exciting the other. To summarize a complex thesis in a few words: *the issue of violence is to this generation what the issue of sex was to the Victorian world.*

The context of development for the post-war generation must again be recalled. These young men and women were born near the end of the most savage, wanton, and destructive war in the history of the world. Perhaps 100,000,000 men, women, and children, most of them "non-combatants," were killed, maimed, or wounded. All of Europe and large parts of Asia and North Africa were laid waste. The lessons of that war for this generation are summarized in the names of three cities: Auschwitz, Hiroshima, Nuremberg. At Auschwitz and the other Nazi concentration camps, more than six million Jews were systematically exterminated. Athough their executioners were sometimes brutal sadists, acts of personal cruelty were the least monentous part of the extermination of European Jewry. Even more impressive are the numbers of "decent," well-educated Germans (who loved their wives, children, and dogs) who learned to take part in, or blind themselves to, this genocide. Murder became depersonalized and dissociated, performed by a System of cold, efficient precision whose members were only following orders in doing a distasteful job well. Bureaucracy, technology, and science were linked in the service of death. Evil became "banal," in Hannah Arendt's words; it was impersonal, dissociated from its human perpetrators, and institutionalized in an efficient and "scientific" organization. It became clear that science and civilization, far from deterring technological death, were its preconditions.

The Second World War ended not with the discovery of the Nazi concentration camps, but with the American use of atomic bombs on the cities of Hiroshima and Nagasaki. This act, which in retrospect hardly seems to have been necessary, helped define the nightmare of the past two decades. Just as the experience of the concentration camp showed that the apparently civilized and "advanced" nations of the world could perform barbarities more cruel than any heretofore imagined, so the atomic bomb and its even more frightening thermonuclear successors provided the concrete imagery for the collective terror of the world. Germany had shown that civilized nations could do the unthinkable; Hiroshima demonstrated how simple, clean, and easy (from the point of view of the perpetrator) doing the unthinkable could be.

In Nuremberg after the war, the German leaders were tried and convicted for their crimes. Here the principle was enunciated and affirmed that there is a law above national interest, an ethic above national purpose, and an accountability above obedience to national leaders. Policies that have the full support of national law may be, nonetheless, criminal and illegal. Confronted with such policies, it is the duty of an ethical man to resist. The principle of collective responsibility was also proposed, and many maintained that the German people, by silence, acquiescence, or deliberate ignorance, had assented to and facillitated the crimes of Nazism.

Auschwitz, Hiroshima, and Nuremberg are the birth pangs of the post-war generation, and their lessons—the bureaucratization of genocide, the clean ease of the unthinkable, and the ethic above nationality—have marked post-modern youth. But despite the nightmare of retaliation that has so far deterred men from the use of their most destructive weapons, the post-war years have not been calm or peaceful. On the contrary, these have been decades of constant international unrest, of continual wars of containment, civil violence, and revolutionary liberation. Since the war, the oppressed two-thirds of the world have largely achieved independence, often through strife, violence, and cruelty. Millions were killed in the victory of the Communist revolution in China; and the struggles for independence in nations like Algeria, Kenya, and Vietnam were cruel and violent. American involvement first in Korea and then in Vietnam, the American "military presence" in dozens of nations across the world, our national policies of "massive retaliation" premised upon city-annihilating thermonuclear weapons, the continually unsuccessful attempt to prevent, limit or control the manufacture of atomic, biological, and chemical weapons—these have been the context for this generation's growth. The marvels of electronic communication have brought these violent realities of the post-war world into the living rooms of almost every young American, concretizing and making emotionally immediate—at least to those who are capable of identification—the absurd violence of the modern world, and the even more frightening possibilities of world-wide cataclysm.

In the lives of young radicals and of their generation, the threat of outer violence has been not only a backdrop, but a constant fact of life. It is reflected not only in childhood terrors of the Bomb, but in the routine experience of air-raid drills in school, in constant exposure to discussions of fallout shelters, preventive warfare, ballistic missiles, and anti-missile defenses, and sometimes in a compulsive fascination with the technology of destruction. The Bomb and what it symbolizes has set the tone for this generation, even for the majority who make a semideliberate point of trying not to think about it. There are relatively few young Americans who, upon hearing a distant explosion, seeing a bright flash of light, or hearing a faraway sound of jets overhead at night, have not wondered for a brief instant whether this might not be "It." And there are a surprising number who have thought, often with horror and dismay, that they wished "It" were over so they would no longer live in fear. Most thoughtful members of the post-war generation have had elaborate fantasies—usually wishful fantasies of survival and rebirth—about what would become of them if "It" happened. All of this points to a great investment of energy, attention, and thought around the issue of violence, although most find the issue too painful to discuss or even to think about.

Continual confrontation with the fact and possibility of violence in the world has activated and become joined with the universal human potential for aggression, anger, and rage: the psychological and historical possibilities of violence have come to strengthen each other. Witnessing the acting out of violence on a scale more gigantic than ever before, or imaginatively participating in holocaust—both activate the fear of inner violence; while heightened awareness of the inner potential for rage, anger, and destructiveness in turn increases sensitivity to violence in the world. It therefore does not require an assumption of increased biological aggression to account for the importance of violence to post-war youth. Starting with the Second World War, we have witnessed violence and imagined violence on a scale more frightening than ever before. Like the angry children in a violent home who fear that their rage will destroy the warring adults around them, we have become vastly more fearful of our inner angers. In fact, we live in a world where even the mildest irritation, multiplied a billionfold by modern technology, might destroy all civilization.

The fear of violence has led to a fascination with it that further surrounds us with its symptoms. Our society is preoccupied with the violence of organized crime, the violence of urban rioting, the violence of an assassinated President and the televised murder of his alleged murderer, the violence of madmen, the oppressed and the rage-filled. And to have been an American child in the past two decades is, as many have noted, to have watched the violence of television, both as it reports the bloodshed of the American and non-American world, and as it skillfully elaborates in repetitive dramas the potential for brutality and aggression in each of us. We have been repeatedly reminded in the past decade that our society, despite its claims to peaceableness and justice, is in fact one of the most violent societies in the history of the world.

In the Victorian era, what was most deeply repressed, rejected, feared, controlled, projected onto others, or compulsively acted out was related to the issue of sex. The personal and social symptomatology of that era—the hysterical ladies who consulted Freud, the repressive moralism of middle-class life, and the sordid underlife of the "other Victorians"—can only be understood in the context of the preoccupation of the Victorian era with human sexuality. The post-war generation, in contrast, is freer, more open, less guilt- and anxiety-ridden about sex. Sex obviously remains important, as befits one of the primary human drives. But increasing numbers of post-modern youth, like these young radicals, have been able to overcome even the asceticism and puritanism of their own adolescences and to move toward a sexuality that is less obsessional, less dissociated, less driven, more integrated with other human experiences and relationships. Inner and outer violence is replacing sex as a prime object of fear, terror, projection, displacement, repression, suppression, acting out, and efforts at control.

At the same time, the symptomatology of violence and repressed violence is becoming more visible. In the complex and highly organized modern world, open displays of rage, anger, and fury are increasingly tabooed: they are considered "irrational"; they threaten to disrupt the finely tuned system in which we live out our working lives; we consider them "childish" or "dangerous." Driven underground, our inevitable angers sometimes seek less direct forms of expression: they heighten autonomic activity to the point of psychosomatic illness; they are turned against the self, producing angry depression; and they are expressed interpersonally in subtle undercutting, backbiting, viciousness, and pettiness. The repression of inner violence makes us eager consumers of the packaged violence of television and the trashy novel. Equally important, our suppressed aggression is projected onto others. We grossly exaggerate the violence of the oppressed, of our enemies, and even of those to whom our society has given good grounds for anger. Consider, for example, the white fear of black violence. Until the summers of 1966 and 1967, it was the Negroes and their friends whose churches were bombed, who were shot, beaten, and injured by whites, and rarely, if ever, vice versa. And even in the urban rioting of 1966 and 1967, the number of black men killed by white men far outweighed the toll of whites. Yet it is the fear of *black* violence that preoccupies the white public.

To connect the fear of black violence, or the war in Vietnam, or the assassination of a President, or the violence of television solely to the threat of technological death would be a gross over simplification. My argument here is simply that we live in an unusually violent society, in an unprecedentedly violent world. In our society—as in others—the fears and facts of violence are self-stimulating. The greater the outer reality of violence, the more the inner fear of it, and for many, the greater the need to create or find external situations in which violence can be experienced vicariously outside themselves. The way men react to constant confrontation with violence in the world of course differs: some tolerate it; others develop symptoms related to their inability to confront directly their own inner anger; others project their inner aggression onto others; still others develop a "neurotic" preoccupation with the possibilities of world holocaust. And, most dangerous of all, some need to act violently in order to discharge their own excited rage. If we are to choose one issue as central to our own time, one danger as most frightening, one possibility as most to be avoided and yet most fascinating, one psychological issue that both rationally and irrationally preoccupies us, it is the issue of violence.

In tracing the development of the young radicals who led Vietnam Summer, I have underlined the related themes that center on the concept of violence. Many of their earliest memories involve conflict, outer anger, and inner fear. They were, throughout their childhoods, especially sensitive to the issue of struggle within their families and communities. Although in behavior most of these young radicals were rather *less* violent than their contemporaries, this was not because they were indifferent to the issue, but because their early experience and family values had taught them how to control, modulate, oppose, and avoid violence. Verbal aggression took the place of physical attack. They learned to argue, to compromise, and to make peace when confronted with conflict. So, too, although their adolescent experience was full of inner conflict, they acted on their often violent feelings only during a brief period of indignant rebellion against the inconsistencies of their parents. These young radicals are unusual in their sensitivity to violence, as in their need and ability to oppose it.

I have mentioned the many tensions—psychological, interpersonal, and organizational—that are related to this issue in their work. The avoidance and control of violence, whether in international warfare, political organizations, small groups, or face-to-face personal relations, is a central goal and a key psychological orientation in the New Left. Many of the problems of the Movement are related to the zealous effort to avoid actions and relations in which inner aggression or outer conflict may be evoked. Recall, for example, the extraordinary efforts made to avoid domination within the Movement, the distrust of "totalitarian intimidation," the suppression of leadership lest it lead to manipulation, the avoidance of "flashiness" that might exploit the organized. Remember, too, the deliberate efforts of many of these young men and women to overcome their own angers, their capacity to stay "cool" when provoked, their initial preference for "non-violent" forms of protest, and their largely successful struggle to overcome in themselves any vestige of sadism, cruelty, domination, or power-seeking in human relationships.

I do not mean to suggest that young radicals in particular, or their generation in general, are rage-filled deniers of their own inner angers. On the contrary, amongst these young radicals, exuberance and zest are the rule rather than the exception. Nor are these young radicals incapable of anger and resentment— although they find these emotions easiest to tolerate when, as in their adolescent rebellions, they can be buttressed by a sense of outraged principle. But young radicals, even more than most young men and women of their generation, learned early in their lives the fruitlessness of conflict; and this lesson, in later years, was among the many forces that went into their decision to work for Vietnam Summer.

The position of the psychologically non-violent revolutionary in opposition to a violent world is paradoxical. On the one hand, he seeks to minimize violence, but, on the other, his efforts often elicit violence from others. He works toward a vague vision of a peaceful world, but he must confront more directly than most of his peers the warfare of the world. The frustrations of his work repetitively reawaken his rage, which must continually be redirected into peaceful paths. Combating destructiveness and exploitation in others, his own destructiveness and desire to exploit are inevitably aroused. Furthermore, he is a citizen of a nation whose international policies seem to him only slightly less barbarous than the policies of the Nazis toward the Jews. He has been recently reminded that, with the support of world opinion, the State of Israel executed Adolf Eichmann because of his complicity in the extermination of the Jews (despite his plea that he was only following orders). Rather than be an accomplice in a comparable enterprise, should the radical not move toward the violent resistance that the world would have preferred from Eichmann? For all his efforts to control violence, cataclysm, and sadism, the young radical continually runs the danger of identifying himself with what he seeks to control, and through a militant struggle against violence, creating more violence than he overcomes. The issue of violence is not resolved for these young men and women. Nor can it be.

In the previous article, Professor Kenneth Keniston analysed the youthful radicalism of "the Movement" as a product of unprecedented historical change, of the fears and tensions created by the threat of global war, and of the affluence generated by the Post-industrial Revolution. Some observers have argued that the generation gap has produced more than just political and social radicalism: the Youth Movement has achieved a genuine cultural break-through—a cultural blossoming that is creating fresh new forms of art, music, and personal expression generally. Representing this point of view, jazz critic and journalist Ralph J. Gleason provides a favorable interpretation of youth culture and the styles of music it has produced. Mr. Gleason's essay raises a number of significant questions, however. Do rock music and youth culture generally stand as a total revolt against "Technetronic Society" or are they indeed just one more product of technology—technology in the form of electronic amplifiers, strobe lights, transistor radios, hallucigens, stereo sets, and television? Does the Youth Movement's violent rejection of parental standards—the puritan ethic, money values, conventional morals and dress—constitute a total and fundamental break with the past, or it is merely one more in a series of generational revolts, relatively harmless in nature, such as that of the "flaming youth" of the 1920's Jazz Age?

Ralph J. Gleason on Rock Music*

*Forms and rhythms in music are never changed without
producing changes in the most political
forms and ways.*
Plato said that.

*There's something happenin' here.
What it is ain't exactly clear.
There's a man with a gun over there*

*Source: Ralph J. Gleason, "Like A Rolling Stone," *The American Scholar*, Autumn 1967 pp. 555-563. Reprinted by permission. Copyright © Ralph J. Gleason.

162

tellin' me I've got to beware.
I think it's time we STOP,
children, what's that sound?
Everybody look what's goin' down.
The Buffalo Springfield said that.

For the reality of politics, we must go to the poets, not the politicians.
Norman O. Brown said that.

For the reality of what's happening today in
America, we must go to rock 'n roll, to popular
music.
I said that.

For almost forty years in this country, which has prided itself on individualism, freedom and nonconformity, all popular songs were written alike. They had an eight-bar opening statement, an eight-bar repeat, an eight-bar middle section or bridge, and an eight-bar reprise. Anything that did not fit into that framework was, appropriately enough, called a novelty.

Clothes were basically the same whether a suit was double-breasted or single-breasted, and the only people who wore beards were absentminded professors and Bolshevik bomb throwers. Long hair, which was equated with lack of masculinity—in some sort of subconscious reference to Samson, I suspect—was restricted to painters and poets and classical musicians, hence the term "long-hair music" to mean classical.

Four years ago a specter was haunting Europe, one whose fundamental influence, my intuition tells me, may be just as important, if in another way, as the original of that line. The Beatles, four long-haired Liverpool teen-agers, were busy changing the image of popular music. In less than a year, they invaded the United States and almost totally wiped out the standard Broadway show—Ed Sullivan TV program popular song. No more were we "flying to the moon on gossamer wings," we were now articulating such interesting and, in this mechanistic society, unusual concepts as "Money can't buy me love" and "I want to hold your hand."

"Societies, like individuals, have their moral crises and their spiritual revolutions," R. H. Tawney says in *Religion and the Rise of Capitalism*. And the Beatles appeared ("a great figure rose up from the sea and pointed at me and said 'you're a Beatle with an "a" ' "—Genesis, according to John Lennon). They came at the proper moment of a spiritual cusp—as the martian in Robert Heinlein's *Stranger in a Strange Land* calls a crisis.

Instantly, on those small and sometimes doll-like figures was focused all the rebellion against hypocrisy, all the impudence and irreverence that the youth of that moment was feeling vis-a-vis his elders.

Automation, affluence, the totality of instant communication, the technology of the phonograph record, the transistor radio, had revolutionized life for youth in this society. The population age was lowering. Popular music, the jukebox and the radio were becoming the means of communication. Huntley and Brinkley were for mom and dad. People now sang songs they wrote themselves, not songs written *for* them by hacks in grimy Tin Pan Alley offices.

The folk music boom paved the way. Bob Dylan's poetic polemics, "Blowin' in the Wind" and "The Times They Are A-Changin'," had helped the breakthrough. "Top-40" radio made Negro music available everywhere to a greater degree than ever before in our history.

This was, truly, a new generation—the first in America raised with music constantly in its ear, weaned on a transistor radio, involved with songs from its earliest moment of memory.

Music means more to this generation than it did even to its dancing parents in the big-band swing era of Benny Goodman. It's natural, then, that self-expression should find popular music so attractive.

The dance of the swing era, of the big bands, was the fox-trot. It was really a formal dance extended in variation only by experts. The swing era's parents had danced the waltz. The fox-trot was a ritual with only a little more room for self-expression. Rock 'n roll brought with it not only the voices of youth singing their protests, their hopes and their expectations (along with their pathos and their sentimentality and their personal affairs from drag racing to romance), it brought their dances.

"Every period which abounded in folk songs has, by the same token, been deeply stirred by Dionysiac currents," Nietzsche points out in *The Birth of Tragedy*. And Dionysiac is the word to describe the dances of the past ten years, call them by whatever name from bop to Twist to the Frug, from the Hully Gully to the Philly Dog.

In general, adult society left the youth alone, prey to the corruption the adults suspected was forthcoming from the song lyrics ("All of me, why not take all of me," from that hit of the thirties, of course, didn't mean *all* of me, it meant, well . . . er . . .) or from the payola-influenced disc jockeys. (Who ever remembers about the General Electric scandals of the fifties, in which over a dozen officials went to jail for industrial illegalities?)

The TV shows were in the afternoon anyway and nobody could stand to watch those rock 'n roll singers; they were worse than Elvis Presley.

But all of a sudden the *New Yorker* joke about the married couple dreamily remarking, when a disc jockey played "Houn' Dog" by Elvis, "they're playing our song," wasn't a joke any longer. It was real. That generation had suddenly grown up and married and Elvis was real memories of real romance and not just kid stuff.

All of a sudden, the world of music, which is big business in a very real way, took another look at the music of the ponytail and chewing gum set, as Mitch

Miller once called the teenage market, and realized that there was one helluva lot of bread to be made there.

In a short few years, Columbia and R.C.A. Victor and the other companies that dominated the recording market, the huge publishing houses that copyrighted the music and collected the royalties, discovered that they no longer were "kings of the hill." Instead, a lot of small companies, like Atlantic and Chess and Imperial and others, had hits by people the major record companies didn't even know, singing songs written in Nashville and Detroit and Los Angeles and Chicago and sometimes, but no longer almost always, New York.

It's taken the big ones a few years to recoup from that. First they called the music trash and the lyrics dirty. When that didn't work, as the attempt more recently to inhibit songs with supposed psychedelic or marijuana references has failed, they capitulated. They joined up. R.C.A. Victor bought Elvis from the original company he recorded for—Sun Records ("Yaller Sun records from Nashville" as John Sebastian sings it in "Nashville Cats")—and then bought Sam Cooke, and A.B.C. Paramount bought Ray Charles and then Fats Domino. And Columbia, thinking it had a baby folk singer capable of some more sales of "San Francisco Bay," turned out to have a tiny demon of a poet named Bob Dylan.

So the stage was set for the Beatles to take over—"with this ring I can—dare I say it?—rule the world!" And they did take over so thoroughly that they have become the biggest success in the history of show business, the first attraction ever to have a coast-to-coast tour in this country sold out before the first show even opened.

With the Beatles and Dylan running tandem, two things seem to me to have been happening. The early Beatles were at one and the same time a declaration in favor of love and of life, an exuberant paean to the sheer joy of living, and a validation of the importance of American Negro music.

Dylan, by his political, issue-oriented broadsides first and then by his Rimbaudish nightmare visions of the real state of the nation, his bittersweet love songs and his pure imagery, did what the jazz and poetry people of the fifties had wanted to do—he took poetry out of the classroom and out of the hands of the professors and put it right out there in the streets for everyone.

I dare say that with the inspiration of the Beatles and Dylan we have more poetry being produced and more poets being made than ever before in the history of the world. Dr. Malvina Reynolds—the composer of "Little Boxes"— thinks nothing like this has happened since Elizabethan times. I suspect even that is too timid an assessment.

Let's go back to Plato, again. Speaking of the importance of new styles of music, he said, "The new style quietly insinuates itself into manners and customs and from there it issues a greater force . . . goes on to attack laws and constitutions, displaying the utmost impudence, until it ends by overthrowing everything, both in public and in private."

That seems to me to be a pretty good summation of the answer to the British rock singer Donovan's question, "What goes on? I really want to know."

The most immediate apparant change instituted by the new music is a new way of looking at things. We see it evidenced all around us. The old ways are going and a new set of assumptions is beginning to be worked out. I cannot even begin to codify them. Perhaps it's much too soon to do so. But I think there are some clues—the sacred importance of love and truth and beauty and interpersonal relationships.

When Bob Dylan sang recently at the Masonic Memorial Auditorium in San Francisco, at intermission there were a few very young people in the corridor backstage. One of them was a longhaired, poncho-wearing girl of about thirteen. Dylan's road manager, a slender, long-haired, "Bonnie Prince Charlie" youth, wearing black jeans and Beatle boots, came out of the dressing room and said, "You kids have to leave! You can't be backstage here!"

"Who are you?" the long-haired girl asked.

"I'm a cop," Dylan's road manager said aggressively.

The girl looked at him for a long moment and then drawled, "Whaaaat? With those boots?"

Clothes really do *not* make the man. But sometimes . . .

I submit that was an important incident, something that could never have happened a year before, something that implies a very great deal about the effect of the new style, which has quietly (or not so quietly, depending on your view of electric guitars) insinuated itself into manners and customs.

Among the effects of "what's goin' on" is the relinquishing of belief in the sacredness of logic. "I was a prisoner of logic and I still am," Malvina Reynolds admits, but then goes on to praise the new music. And the prisoners of logic are the ones who are really suffering most—unless they have Mrs. Reynolds' glorious gift of youthful vision.

The first manifestation of the importance of this outside the music—I think—came in the works of Ken Kesey and Joseph Heller. *One Flew Over the Cuckoo's Nest,* with its dramatic view of the interchangeability of reality and illusion, and *Catch-22,* with its delightful utilization of crackpot realism (to use C. Wright Mills's phrase) as an explanation of how things are, were works of seminal importance.

No one any longer really believes that the processes of international relations and world economics are rationally explicable. Absolutely the very best and clearest discussion of the entire thing is wrapped up in Milo Minderbinder's explanation, in *Catch-22,* of how you can buy eggs for seven cents apiece in Malta and sell them for five cents in Pianosa and make a profit. Youth understands the truth of this immediately, and no economics textbook is going to change it.

Just as—implying the importance of interpersonal relations and the beauty of being true to oneself—the under-thirty youth immediately understands the creed patiently explained by Yossarian in *Catch-22* that everybody's your enemy who's trying to get you killed, even if he's your own commanding officer.

This is an irrational world, despite the brilliant efforts of Walter Lippmann to make it rational, and we are living in a continuation of the formalized lunacy (Nelson Algren's phrase) of war, any war.

At this point in history, most of the organs of opinion, from the *New York Review of Books* through the *New Republic* to *Encounter* (whether or not they are subsidized by the C.I.A.), are in the control of the prisoners of logic. They take a flick like *Morgan* and grapple with it. They take *Help* and *A Hard Day's Night* and grapple with those two beautiful creations, and they fail utterly to understand what is going on because they try to deal with them logically. They complain because art doesn't make sense! Life on this planet in this time of history doesn't make sense either—as an end result of immutable laws of economics and logic and philosophy.

Dylan sang, "You raise up your head and you ask 'is this where it is?' And somebody points to you and says 'it's his' and you say 'what's mine' and somebody else says 'well, what is' and you say 'oh my god am i here all alone?' "

Dylan wasn't the first. Orwell saw some of it. Heller saw more, and in a different way so did I. F. Stone, that remarkable journalist, who is really a poet, when he described a *Herald Tribune* reporter extracting from the Pentagon the admission that, once the first steps for the Santo Domingo episode were mounted, it was impossible to stop the machine.

Catch-22 said that in order to be sent home from flying missions you had to be crazy, and obviously anybody who wanted to be sent home was sane.

Kesey and Heller and Terry Southern, to a lesser degree in his novels but certainly in *Dr. Strangelove,* have hold of it. I suspect that they are not really a *New Wave* of writers but only a *last* wave of the past, just as is Norman Mailer, who said in his Berkeley Vietnam Day speech that "rational discussion of the United States' involvement in Viet Nam is illogical in the way surrealism is illogical and rational political discussion of Adolf Hitler's motives was illogical and then obscene." This is the end of the formal literature we have known and the beginning, possibly, of something else.

In almost every aspect of what is happening today, this turning away from the old patterns is making itself manifest. As the formal structure of the show business world of popular music and television has brought out into the open the Negro performer—whose incredibly beautiful folk poetry and music for decades has been the prime mover in American song—we find a curious thing happening.

The Negro perfomers, from James Brown to Aaron Neville to the Supremes and the Four Tops, are on an Ed Sullivan trip, striving as hard as they can to get on that stage and become part of the American success story, while the white rock performers are motivated to escape from that stereotype. Whereas in years past the Negro performer offered style in performance and content in song—the messages from Leadbelly to Percy Mayfield to Ray Charles were important messages—today he is almost totally style with very little content. And when James Brown sings, "It's a Man's World," or Aaron Neville sings, "Tell It Like It Is," he takes a phrase and only a phrase with which to work, and the Supremes and the Tops are choreographed more and more like the Four Lads and the Ames Brothers and the McGuire Sisters.

I suggest that this bears a strong relationship to the condition of the civil rights movement today in which the only truly black position is that of Stokely Carmichael, and in which the N.A.A.C.P. and most of the other formal groups are, like the Four Tops and the Supremes, on an Ed Sullivan-TV-trip to middle-class America. And the only true American Negro music is that which abandons the concepts of European musical thought, abandons the systems of scales and keys and notes, for a music whose roots are in the culture of the colored peoples of the world.

The drive behind all American popular music performers, to a greater or lesser extent, from Sophie Tucker and Al Jolson, on down through Pat Boone and as recently as Roy Head and Charlie Rich, has been to sound like a Negro. The white jazz musician was the epitome of this.

Yet an outstanding characteristic of the new music of rock, certainly in its best artists, is something else altogether. This new generation of musicians is not interested in being Negro, since that is an absurdity.

The clarinetist Milton Mezzrow, who grew up with the Negro Chicago jazzmen in the twenties and thirties, even put "Negro" on his prison record and claimed to be more at home with his Negro friends than with his Jewish family and neighbors.

Today's new youth, beginning with the rock band musician but spreading out into the entire movement, into the Haight-Ashbury hippies, is not ashamed of being white.

He is remarkably free from prejudice, but he is not attempting to join the Negro culture or to become part of it, like his musical predecessor, the jazzman, or like his social predecessor, the beatnik. I find this of considerable significance. For the very first time in decades, as far as I know, something important and new is happening artistically and musically in this society that is distinct from the Negro and to which the Negro will have to come, if he is interested in it at all, as in the past the white youth went uptown to Harlem or downtown or crosstown or to wherever the Negro community was centered because there was the locus of artistic creativity.

Today the new electronic music by the Beatles and others (and the Beatles' "Strawberry Fields" is, I suggest, a three-minute masterpiece, an electronic miniature symphony) exists somewhere else from and independent of the Negro. This is only one of the more easily observed manifestations of this movement.

The professional craft union, the American Federation of Musicians, is now faced with something absolutely unforeseen—the cooperative band. Briefly—in the thirties—there were co-op bands. The original Casa Loma band was one and the original Woody Herman band was another. But the whole attitude of the union and the attitude of the musicians themselves worked against the idea, and co-op bands were discouraged. They were almost unknown until recently.

Today almost all the rock groups are cooperative. Many live together, in tribal style, in houses or camps or sometimes in traveling tepees, but always *together* as a *group;* and the young girls who follow them are called "groupies," just as the girls who in the thirties and forties followed the bands (music does more than soothe the savage beast!) were called "band chicks."

The basic creed of the American Federation of Musicians is that musicians must not play unless paid. The new generation wants money, of course, but its basic motivation is to play anytime, anywhere, anyhow. Art is first, then finance, most of the time. And at least one rock band, the Loading Zone in Berkeley, has stepped outside the American Federation of Musicians entirely and does not play for money. You may give them money, but they won't set a price or solicit it.

This seems to me to extend the attitude that gave Pete Seeger, Joan Baez and Bob Dylan such status. They are not and never have been for sale in the sense that you can hire Dean Martin to appear, any time he's free, as long as you pay his price. You have not been able to do this with Seeger, Baez and Dylan any more than Allen Ginsberg has been for sale either to *Ramparts* or the C.I.A.

Naturally, this revolt against the assumptions of the adult world runs smack dab into the sanctimonious puritan morality of America, the schizophrenia that insists that money is serious business and the acquisition of wealth is a blessing in the eyes of the Lord, that what we do in private we must preach against in public. Don't do what I do, do what I say.

Implicit in the very names of the business organizations that these youths form is an attack on the traditional, serious attitude toward money. It is not only that the groups themselves are named with beautiful imagery: the Grateful Dead, the Loading Zone, Blue Cheer or the Jefferson Airplane—all dating back to the Beatles with an A—it is the names of the nonmusical organizations: Frontage Road Productions (the music company of the Grateful Dead), Faithful Virtue Music (the Lovin' Spoonful's publishing company), Ashes and Sand (Bob Dylan's production firm—his music publishing company is Dwarf Music). A group who give light shows is known as the Love Conspiracy Commune, and there was a dance recently in Marin County, California, sponsored by the Northern California Psychedelic Cattlemen's Association, Ltd. And, of course, there is the Family Dog, which, despite *Ramparts,* was never a rock group, only a name under which four people who wanted to present rock 'n roll dances worked.

Attacking the conventional attitude toward money is considered immoral in the society of our fathers, because money is sacred. The reality of what Bob Dylan says—"money doesn't talk, it swears"—has yet to seep through.

A corollary of the money attack is the whole thing about long hair, bare feet and beards. "Nothing makes me sadder," a woman wrote me objecting to the Haight-Ashbury scene, than to see beautiful girls walking along the street in bare feet." My own daughter pointed out that your feet couldn't get any dirtier than your shoes.

Recently I spent an evening with a lawyer, a brilliant man who is engaged in a lifelong crusade to educate and reform lawyers. He is interested in the civil liberties issue of police harassment of hippies. But, he said, they wear those uniforms of buckskin and fringe and beads. Why don't they dress naturally? So I asked him if he was born in this three-button dacron suit. It's like the newspaper descriptions of Joan Baez's "long stringy hair." It may be long, but *stringy*? Come on!

To the eyes of many of the elder generation, all visible aspects of the new generation, its music, its lights, its clothes, are immoral. The City of San Francisco Commission on Juvenile Delinquency reported adversely on the sound level and the lights at the Fillmore Auditorium, as if those things of and by themselves were threats (they may be, but not in the way the Commission saw them). A young girl might have trouble maintaining her judgment in that environment, the Commission chairman said.

Now this all implies that dancing is the road to moral ruin, that young girls on the dance floor are mesmerized by talent scouts for South American brothels and enticed away from their happy (not hippie) homes to live a life of slavery and moral degradation. It ought to be noted, parenthetically, that a British writer, discussing the Beatles, claims that "the Cycladic fertility goddess from Amorgos dates the guitar as a sex symbol to 4800 years B. C."

During the twenties and the thirties and the forties—in other words, during the prime years of the Old Ones of today—dancing, in the immortal words of Bob Scobey, the Dixieland trumpet player, "was an excuse to get next to a broad." The very least effect of the pill on American youth is that this is no longer true.

The assault on hypocrisy works on many levels. The adult society attempted to chastise Bob Dylan by economic sanction, calling the line in "Rainy Day Woman," "everybody must get stoned" (although there is a purely religious, even biblical, meaning to it, if you wish), an enticement to teen-agers to smoke marijuana. But no one has objected to Ray Charles's "Let's Go Get Stoned," which is about gin, or to any number of other songs, from the Kingston Trio's "Scotch and Soda" on through "One for My Baby and One More [ONE MORE!] for the Road." Those are about alcohol and alcohol is socially acceptable, as well as big business, even though I believe that everyone under thirty now knows that alcohol is worse for you than marijuana, that, in fact, the only thing wrong about marijuana is that it is illegal.

Cut to the California State Narcotics Bureau's chief enforcement officer, Matt O'Connor, in a TV interview recently insisting, à la Parkinson's Law, that he must have more agents to control the drug abuse problem. He appeared with a representative of the state attorney general's office, who predicted that the problem would continue "as long as these people believe they are not doing anything wrong."

And that's exactly it. They do not think they are doing anything wrong, any more than their grandparents were when they broke the prohibition laws. They do not want to go to jail, but a jail sentence or a bust no longer carries the social stigma it once did. The civil rights movement has made a jailing a badge of honor, if you go there for principle, and to a great many people today, the right to smoke marijuana is a principle worth risking jail for.

"Make Love, Not War" is one of the most important slogans of modern times, a statement of life against death, as the Beatles have said over and over—"say the word and be like me, say the word and you'll be free."

I don't think that wearing that slogan on a bumper or on the back of a windbreaker is going to end the bombing tomorrow at noon, but it implies something. It is not conceivable that it could have existed in such proliferation thirty years ago, and in 1937 we were pacifists, too. It simply could not have happened.

There's another side to it, of course, or at least another aspect of it. The Rolling Stones, who came into existence really to fight jazz in the clubs of London, were against the jazz of the integrated world, the integrated world arrived at by rational processes. Their songs, from "Satisfaction" and 19th Nervous Breakdown" to "Get Off of My Cloud" and "Mother's Little Helper," were antiestablishment songs in a nonpolitical sort of way, just as Dylan's first period was antiestablishment in a political way. The Stones are now moving, with "Ruby Tuesday" and "Let's Spend the Night Together," into a social radicalism of sorts; but in the beginning, and for their basic first-thrust appeal, they hit out in rage, almost in blind anger and certainly with overtones of destructiveness, against the adult world. It's no wonder the novel they were attracted to was David Wallis' *Only Lovers Left Alive,* that Hell's Angels story of a teen-age, future jungle. And it is further interesting that their manager, Andrew Loog Oldham, writes the essays on their albums in the style of Anthony Burgess' violent *A Clockwork Orange.*

Nor is it any wonder that this attitude appealed to that section of the youth whose basic position was still in politics and economics (remember that the Rolling Stone Mick Jagger was a London School of Economics student, whereas Lennon and McCartney were artists and writers). When the Stones first came to the West Coast, a group of young radicals issued the following proclamation of welcome:

> Greetings and welcome Rolling Stones, our comrades in the desperate battle against the maniacs who hold power. The revolutionary youth of the world hears your music and is inspired to even more deadly acts. We fight in guerrilla bands against the invading imperialists in Asia and South America, we riot at rock n' roll concerts everywhere. We burned and pillaged in Los Angeles and the cops know our snipers will return.
>
> They call us dropouts and delinquents and draftdodgers and punks and hopheads and heap tons of shit on our heads. In Viet Nam they drop bombs on us and in America they try to make us make war on our own comrades but the bastards hear us playing you on our little transistor radios and know that they will not escape the blood and fire of the anarchist revolution.
>
> We will play your music in rock 'n roll marching bands as we tear down the jails and free the prisoners, as we tear down the State schools and free the students, as we tear down the military bases and arm the poor, as we tatoo BURN BABY BURN! on the bellies of the wardens and generals and create a new society from the ashes of our fires.
>
> Comrades, you will return to this country when it is free from the tyranny of the State and you will play your splendid music in factories run by the workers, in the domes of emptied city halls, on the rubble of police stations, under the hanging corpses of priests, under a million red flags waving over a million anarchist communities. In the words of Breton, THE ROLLING STONES ARE THAT WHICH SHALL BE! LYNDON JOHNSON—THE YOUTH OF CALIFORNIA DEDICATES ITSELF TO YOUR DESTRUCTION! ROLLING STONES—THE YOUTH OF CALIFORNIA HEARS YOUR MESSAGE! LONG LIVE THE REVOLUTION!!!

But rhetoric like that did not bring out last January to a Human Be-In on the polo grounds of San Francisco's Golden Gate Park twenty thousand people who were there, fundamentally, just to see the other members of the tribe, not to hear speeches—the speeches were all a drag from Leary to Rubin to Buddah*—but just to BE.

In the Haight-Ashbury district the Love Generation organizes itself into Job Co-ops and committees to clean the streets, and the monks of the neighborhood, the Diggers, talk about free dances in the park to put the Avalon Ballroom and

---

*The Be-In heard speeches by Timothy Leary, the psychedelic guru, Jerry Rubin, the leader of the Berkeley Vietnam Day movement, and Buddah, a bartender and minor figure in the San Francisco hippie movement who acted as master of ceremonies.

the Fillmore out of business and about communizing the incomes of Bob Dylan and the Beatles.

The Diggers trace back spiritually to those British millenarians who took over land in 1649, just before Cromwell, and after the Civil War freed it, under the assumption that the land was for the people. They tilled it and gave the food away.

The Diggers give food away. Everything is Free. So is it with the Berkeley Provos and the new group in Cleveland—the Prunes—and the Provos in Los Angeles. More, if an extreme, assault against the money culture. Are they driving the money changers out of the temple? Perhaps. The Diggers say they believe it is just as futile to fight the system as to join it and they are dropping out in a way that differs from Leary's.

The Square Left wrestles with the problem. They want a Yellow Submarine community because that is where the strength so obviously is. But even *Ramparts,* which is the white hope of Square Left, if you follow me, misunderstands. They think that the Family Dog is a rock group and that political activity is the only hope, and Bob Dylan says, "There's no left wing and no right wing, only up wing and down wing," and also, "I tell you there are no politics."

But the banding together to form Co-ops, to publish newspapers, to talk to the police (even to bring them flowers), aren't these political acts? I suppose so, but I think they are political acts of a different kind, a kind that results in the Hell's Angels being the guardians of the lost children at the Be-In and the guarantors of peace at dances.

The New Youth is finding its prophets in strange places—in dance halls and on the jukebox. It is on, perhaps, a frontier buckskin trip after a decade of Matt Dillon and Bonanza and the other TV folk myths, in which the values are clear (as opposed to those in the world around us) and right is right and wrong is wrong. The Negro singers have brought the style and the manner of the Negro gospel preacher to popular music, just as they brought the rhythms and the feeling of the gospel music, and now the radio is the church and Everyman carries his own walkie-talkie to God in his transistor.

Examine the outcry against the Beatles for John Lennon's remark about being more popular than Jesus. No radio station that depended on rock 'n roll music for its audience banned Beatles records, and in the only instance where we had a precise measuring rod for the contest—the Beatles concert in Memphis where a revival meeting ran day and date with them—the Beatles won overwhelmingly. Something like eight to five over Jesus in attendance, even though the Beatles charged a stiff price and the Gospel according to the revival preacher was free. Was my friend so wrong who said that if Hitler were alive today, the German girls wouldn't allow him to bomb London if the Beatles were there?

"Nobody ever taught you how to live out in the streets," Bob Dylan sings in "Like a Rolling Stone." You may consider that directed at a specific person, or you may, as I do, consider it poetically aimed at plastic uptight America, to use a phrase from one of the Family Dog founders.

"Nowhere to run, nowhere to hide," Martha and the Vandellas sing, and Simon and Garfunkel say, "The words of the prophets are written on the subway walls, in tenement halls." And the Byrds sing, "A time for peace, I swear it's not too late," just as the Beatles sing, "Say the word." What has formal religion done in this century to get the youth of the world so well acquainted with a verse from the Bible?

Even in those artists of the second echelon who are not, like Dylan and the Beatles and the Stones, worldwide in their influence, we find it. "Don't You Want Somebody To Love," the Jefferson Airplane sings, and Bob Lind speaks of "the bright elusive butterfly of love."

These songs speak to us in our condition, just as Dylan did with "lookout kid, it's somethin' you did, god knows what, but you're doin' it again." And Dylan sings again a concept that finds immediate response in the tolerance and the antijudgment stance of the new generation, when he says, "There are no trials inside the Gates of Eden."

Youth is wise today. Lenny Bruce claimed that TV made even eight-year-old girls sophisticated. When Bob Dylan in "Desolation Row" sings, "At midnight all the agents and the superhuman crew come out and round up everyone that knows more than they do," he speaks true, as he did with "don't follow leaders." But sometimes it is, as John Sebastian of the Lovin' Spoonful says, "like trying to tell a stranger 'bout a rock 'n roll."

Let's go back again to Nietzsche.

> Orgiastic movements of a society leave their traces in music [he wrote]. Dionysiac stirrings arise either through the influence of those narcotic potions of which all primitive races speak in their hymns [–dig that!–] or through the powerful approach of spring, which penetrates with joy the whole frame of nature. So stirred, the individual forgets himself completely. It is the same Dionysiac power which in medieval Germany drove ever increasing crowds of people singing and dancing from place to place; we recognize in these St. John's and St. Vitus' dancers the bacchic choruses of the Greeks, who had their precursors in Asia Minor and as far back as Babylon and the orgiastic Sacea. There are people who, either from lack of experience or out of sheer stupidity, turn away from such phenomena, and strong, in the sense of their own sanity, label them either mockingly or pityingly "endemic diseases." These benighted souls have no idea how cadaverous and ghostly their "sanity" appears as the intense throng of Dionysiac revelers sweeps past them.

And Nietzsche never heard of the San Francisco Commission on Juvenile Delinquency or the Fillmore and the Avalon ballrooms.

"Believe in the magic, it will set you free," the Lovin' Spoonful sing. "This is an invitation across the nation," sing Martha and the Vandellas, and the Mamas and the Papas, "a chance for folks to meet, there'll be laughin', singin' and music swingin', and dancin' in the street!"

Do I project too much? Again, to Nietzsche. "Man now expresses himself through song and dance as the member of a higher community; he has forgotten how to walk, how to speak and is on the brink of taking wing as he dances . . . no longer the *artist,* he has himself become *a work of art.*"

"Hail hail rock 'n roll," as Chuck Berry sings. "Deliver me from the days of old!"

I think he's about to be granted his wish.

One of the most colorful figures in the debate over the significance of the Technological Revolution is R. Buckminster Fuller: inventor of the geodesic dome, philosopher, and spellbinding lecturer. Much like his fellow septagenarian, Lewis Mumford, Fuller is a generalist, a "big picture" man, who surveys man's problems and man's potential on a global or even galactic scale. Emphatically *unlike* Mumford, Buckminster Fuller embraces technology as the one true source of mankind's salvation. He demonstrates this philosophy in the following excerpt. If the Youth Movement really wants to change society, argues Fuller, it must realize that there is only one way to do it—via technology.

R. Buckminster Fuller on Remaking the World*

*   *   *   *   *

Logically, today's youth becomes exasperated and asks: "Why can't we make the world work?" All this negative nonsense is the consequence of outworn, ignorant biases of old timers. I say let's join forces and set things to rights. Parading in multitudes, students demand that their political leaders take steps to bring about peace and plenty. The fallacy of this lies in their age-old and mistaken assumption that the problem is one of political reform. The fact is that the politicians are faced with a vacuum, and you can't reform a vacuum. The vacuum is the apparent world condition of not enough to go around, not enough for even a majority of mankind to survive more than half of its potential life span. It is again a you-or-me-to-the-death situation that leads from impasse to impasse to ultimate showdown by arms. Thus, more and more students around the world are learning of the new and surprising alternative to politics—the design-science revolution that alone can solve the problem.

•Source: From "Education for Comprehensivity," by R. Buckminster Fuller, published in *Approaching the Benign Environment,* edited by Taylor Littleton pp. 70-77. Copyright © 1970 by the University of Alabama Press. (Sections omitted with the permission of the Publisher.)

Along with the 1965 Berkeley episodes and the myriad of civil disturbances which occurred in the United States as a result of the national reactionary pullback at the by-election polls in 1966, every college campus has had its quota of incidents. But in relation to this there are now new realizations. The students who graduated in June, 1966 were born and grew up under historically unprecedented conditions that are known by behavioral science research to have affected very seriously the intellectual and social development of humans. The university and college classes of 1966 were born as the first atomic bomb exploded over Hiroshima, killing and maiming vast numbers of civilians. The students of 1966 were the first human beings to be reared by the third parent—by television, whose voice and presence are seen and heard by children far more than those of the two blood parents. The real parents come home from the store, or office, or golf course, or hairdresser and say: "Wow! what a day! Let's have a beer," and sit down to small talk about local events—and the children slip off to hear the third parent brief them visually, ergo vividly, on the world-around news regarding the world's continual aches, pains, disasters, olympic triumphs, and all.

As the class of 1966 grew and developed from birth, they learned from the third parent about man's going across the North Pole under ice. In their fourteenth year, the Russian unmanned rocket photographed the far side of the moon and returned to earth. When they were fifteen, the U.S. bathyscape took man safely to photograph the bottom of the Pacific Ocean's deepest hole. In their sixteenth year, a Russian orbited earth in a rocket. As they reached seventeen, the DNA genetic code for control of the design of all life was discovered. And then the class of 1966 at Berkeley shocked the world by saying that it felt no special loyalty to its families, its university, its state, or its nation. The youth of the class of 1966 were thought by most oldsters to be shockingly "immoral" and lacking in idealism. Not so. They are as idealistic and full of compassion as any child has ever been, but their loyalty is to all humanity. They are no longer the creatures of local, class, or race biases. They say: "Mankind can do anything it wants. Why don't our officials and families stop talking about their local biases and wasting wealth on warring—all because they assume that war is necessary simply because there does not seem to be enough to take care of even one half of humanity's needs." Today's youth say: "Why not up the performance per unit of invested resources and thus make enough to go around?" And they're right.

Students can *learn* the following: that technical evolution has this fundamental behavior pattern. First, as I have explained to you, there is a scientific discovery of a generalized principle, which occurs as a subjective realization by experimentally probing man. Next comes objective employment of that principle in a special-case invention. Next, the invention is reduced to practice. This gives man an increased technical advantage over his physical environment. If successful as a tool of society, it is used in ever bigger, swifter, and everyday ways. For instance, it goes progressively from a little steel steamship to ever bigger fleets of constantly swifter, high-powered ocean giants.

There comes a time, however, when we discover other ways of doing the same task more economically. For instance, we discover that a 200-ton transoceanic jet airplane, considered on an annual round-trip frequency basis, can outperform the passenger-carrying capability of the 85,000 ton *Queen Elizabeth* or that a quarter-ton transoceanic communications relay satellite outperforms 150,000 tons of transoceanic cables. All the technical curves rise in tonnage and volumetric size to reach a giant peak, after which progressive miniaturization sets in. After that, a new and more economical art takes over which also goes through the same cycle of doing progressively more with less. First, by getting bigger and taking advantage, for instance, of the fact that doubling the length of a ship increases its wetted surface fourfold while increasing its payload volume eightfold. Inasmuch as the cost of driving progressively bigger ships through the water at a given speed increases in direct proportion to the increase in friction of the wetted surface, the eightfolding of payload volume gained with each fourfolding of wetted surface means twice as much profit for less effort each time the ship's length is doubled. This principle of advantage gain through geometrical-size increase holds true for ships of both air and water. Then doubling of length of sea-going ships finally runs into trouble; for instance, the ocean liner made more than a thousand feet long would have to span between two giant waves and would have to be doubled in size to do so. However, if doubled in size once more, she could no longer be accommodated by the great world canals, dry docks, and harbors.

At this point the miniaturization of doing more with less first ensues through substitution of an entirely new art. David's sling-stone over Goliath's club operated from beyond reach of the giant. This overall and inexorable trending to do more with less is known sum totally as progressive ephemeralization. Ephemeralization trends toward an ultimate doing of everything with nothing at all—which is a trend of the omniweighable physical to be mastered by the omniweightless metaphysics of intellect. All the ballistic arts of man and men's warring to the death have followed this same fundamental evolutionary pattern of bigger, then smaller.

Assuming that there was not and never would be enough of the vital support resources to go around, and concluding that there must be repeating eventualities in wars to see which side could pursue its most favored theory of survival under fundamental inadequacies, humanity has continually done more killing with less human effort at greater and greater distances and at ever higher speeds with ever increasing accuracy. Humanity's killing capability has gone from a thrown stone to a spear to a sling to a bow and arrow to a pistol, a musket, a cannon, and so on to the great weapons-carrying battleships. Then suddenly a little two-ton torpedo-carrying airplane sank a 45,000-ton battleship, and then the 2,000 miles-per-hour airplane was out-performed by the 16,000 miles-per-hour atom-bomb-carrying rocket, a miniscule weight in comparison to the bomb-carrying plane. If world-warring persists as a consequence of the concept of survival only of the fittest minority, there will come the virtually weightless death-rays operating at 700-million miles per hour. At the present point of history, the uranium bomb has been recently displaced by the hydrogen bomb until it was discovered that if either side used that new greatest weapon, both sides and the rest of humanity would perish. Therefore, the biggest weapons could not be used, nor could the equally large and mutually destructive biological or chemical gas warfaring. Then both sides discovered that killing of the enemy's people was not their objective. Killing the enemy's *ideology* is the objective. Killing the enemy's people brings sympathy and support for the enemy from the rest of the world, and gaining the good opinion and support of the rest of the world is one of the new world's war aims.

And now? Both sides have started to explore the waging of more war with lesser, more limited killings, but more politically and economically devastating techniques. Just as ephemeralization (employing ever more miniscule instruments) took technology out of the limited ranges of the human senses into the vast and invisible ranges of the electromagnetic spectrum, so too has major warfaring almost disappeared from the visible contacts of human soldiery and entered into the realm of invisible psychology. In the new invisible miniaturization phase of major world warring, both sides carry on an attention-focusing guerilla warfare, as now in Vietnam, while making their most powerful attacks through subversion, vandalism, or skillful agitation of any and all possible areas of discontent within the formally assumed enemy's home economics.

In carrying on this new and unfamiliar world warring, they don't have to send ideological proselytizers to persuade the people of the other side to abandon their home country's political systems and adopt that of their former enemy. Instead they can readily involve, induce, and persuade individuals of the other side to look for discontent wherever it manifests itself and thereafter to amplify that condition by whatever psychological means until the situation erupts in public demonstrations, etc. The idea is to make a mess of the other's economy and thereby discredit his socio-political system in the eyes of the rest of the world and to destroy the enemy people's confidence in its own system. Because the active operators are most often engaged on the basis of just gratifying their own personal discontent, they are most often unaware that they are acting as agents. Because almost everyone has at least one discontent, a single well-trained, conscious agent can evoke the effective but unwitting agency of hundreds of other discontents, promoters, and joiners. As a consequence of this new invisible phase of world war trending, a most paradoxical condition exists wherein the highly idealistic youth of college age, convinced that they are demonstrating against war, are in fact the front line soldiers operating as unwitting shock troops while the conventionally recognized soldiers, engaged in visible war-zone warfare either of ambush or open battle, are carrying on only a secondary albeit often mortally fatal, decoy operation.

This invisible world-around warring to destroy the enemy's economy wherever it is operative, above all by demonstrating its homeland weaknesses and vulnerabilities to the rest of the world and thus hopefully destroying the confidence of the enemy people in themselves, is far more devastating than a physical death ray could be; for it does everything with nothing. Furthermore, it operates as news that moves around the earth by electromagnetic waves operating at 700-million miles per hour. At the moment, the highly controlled political state has a great defensive advantage over the open, freedom-nurturing states by virtue of the former's controlled news, for it is the omniexcitable news in the free countries which is primarily exploited to publish, spread, and thus create a chain reaction of dismay through the guaranteed publishing of any and all of their organized-discontent activities.

While all the foregoing curves of rising and falling in the technical evolution of weaponry have taken place, there has also occurred, all unnoticed to the parties of the warfaring, a vast fallout from the defense technology into the domestic technology of ephemeralization's doing ever more with ever less. As I remarked, within two-thirds of a century this unnoticed and inadvertent fallout has converted forty-plus percent of total humanity from havenotness to a high standard of living-haveness and made clear that the only way all humanity may be elevated to such an advantage is by further acceleration of this technological revolution.

It becomes evident, then, that youth's world-around clamor for peace can only be realized through technological revolution, which will do so much more with so much less per each function as ultimately to produce enough to support all humanity. It is also clear that such a task can *only* be accomplished by this technological design revolution. As many who have become involved in the new invisible warfaring discover that their aim can only be attained through this design revolution, all the young world-around idealists will have to face the question of whether they prefer to keep on agitating simply because they have come to enjoy a sense of power and importance by doing so, or whether they really are dedicated to the earliest possible attainment of economic and physical success for all humanity and thereby realistically to eliminate war.

If the desire for the elimination of war is what they are most moved by, they will have to shift their effort from mere political agitation to participation in the design science revolution. The latter course involves the development of ever self-regenerating and improving scientific and technical competence, and that in turn means the individual must plunge earnestly and dedicatedly into self-development by the resources of an educational system designed to develop the inherent comprehensivity of humanity.

To social critic Lewis Mumford, whose work has been cited earlier in this book, the most frightening aspect of today's world is that technology has been allowed to take control, to organize society itself into a huge, dehumanized "megamachine." Architecture, art, music, scholarship, and human culture generally have been debased by this perverted, militarized, technocratic system. Thus, Mumford believes that the Youth Revolution's rejection of the values of contemporary society was instinctively correct. The *form* which this revolt has taken, however, does not receive Mumford's approval. The youth, in their very rebellion, he argues, have fallen victim to the very same technological complex they are reacting against—the "System" still has them under its subtle control.

Lewis Mumford on the Mass Mobilization of Youth*

Despite the well-founded dissatisfaction of the younger generation with the kind of life offered by the bloated affluence of megatechnic society, their very mode of rebellion too often demonstrates that the power system still has them in its grip: they, too, mistake indolence for leisure and irresponsibility for liberation. The so-called Woodstock Festival was no spontaneous manifestation of joyous youth, but a strictly money-making enterprise, shrewdly calculated to exploit their rebellions, their adulations, and their illusions. The success of the festival was based on the tropismic attraction of 'Big Name' singers and groups (the counter-culture's Personality Cult!), idols who command colossal financial rewards from personal appearances and the sales of their discs and films.

With its mass mobilization of private cars and buses, its congestion of traffic en route, and its large scale pollution of the environment, the Woodstock Festival mirrored and even grossly magnified the worst features of the system that many young rebels profess to reject, if not to destroy. The one positive achievement of this mass moblization, apparently, was the warm sense of instant

---

*Source: From *THE MYTH OF THE MACHINE: THE PENTAGON OF POWER*, copyright © 1964, 1970 by Lewis Mumford. Reprinted by permission of Harcourt Brace Jovanovich, Inc.

fellowship produced by the close physical contact of a hundred thousand bodies floating in the haze and daze of pot. Our present mass-minded, over-regimented, depersonalized culture has nothing to fear from this kind of reaction—equally regimented, equally depersonalized, equally under external control. What is this but the Negative Power Complex, attached by invisible electrodes to the same pecuniary pleasure center?

What Have They Done to the Rain?
Just a little boy standing in the rain,
The gentle rain that falls for years,
And the grass is gone, the boy disappears,
And the rain keeps falling like helpless tears,
And what have they done to the rain?

Malvina Reynolds●

V. Technology and the Global Environment

The transition from the previous readings on the interrelationship of technology and the student rebellion to the next topic—ecology—is facilitated by an essay entitled, "A Student Manifesto on the Environment." It is significant that the author, Mr. Pennfield Jensen, has combined a deep commitment to the urban poor with his interest in nature and ecological studies. He is Executive Director of UNIFY, the Urban Nature Institute for Youth. Mr. Jensen's passionate concern for the fate of "planet earth" would seem also to reflect an equally profound revulsion against science and technology—in fact, against the entire apparatus of a highly developed industrial society.

Pennfield Jensen on
A Student Manifesto on
the Environment*

The phenomenon of student activism is as much a barometer of global crises as it is a manifestation of personal frustration and organized disruption. The celebrated generation gap is little more than the naturally holistic consciousness of young people facing a way of life that is not only ugly, irrelevant, and neurotic but that threatens to destroy us all. The natural environment, on the other hand, presents to the sensually connected but culturally shocked young person the clear light of moral value and societal obligation. Earth: Love it or leave it.

The impatience demonstrated with the establishment is the best part of today's activism. The worst part is seldom seen for what it really is: a despairing apathy that stultifies all endeavor. The activist is basically a constructivist, a creative and productive person dedicated to "making it better" while, at the same time, demonstrating that the culture, the economics, and the politics of the United States are hopelessly antediluvian. It's not right. It's not working. Shut it down. The healthy concerns of today are directed toward the environment and reach beyond all national boundaries. For nationalism itself is a disease of the mind that settles over a country, smothering its intelligence under a blanket of rot thicker than the smog we breathe. When a young man's life becomes shattered by the blind trauma of a useless war or by the faceless sadism behind an official load of buckshot, one hears windows begin to break the world over. These are dead-ends. Ultimately, activism wants a big answer to a big question. We don't want merely to survive; we want to live. There is only one place to live and that is on this planet and we must live here together.

While individuals of stature and wisdom are arguing for an international ecological congress to establish laws for international use of the earth's resources, the ecological crisis has already precipitated student activism into one of the world's most potentially constructive forces. The activists do not struggle against educational systems because education is despised but because education is needed. The naïveté, enthusiasm, and idealism of young people is not a thing to be scorned; it is rather to be celebrated as the raw material of constructive growth.

•Source: Pennfield Jensen, "A Student Manifesto on the Environment," *Natural History*, April, 1970 pp. 20-22. Reprinted by permission of *Natural History*.

The ecological perspective shows all of life connected into dynamic processes with ineluctable consequences should those processes be changed.

The ecological sentence for mankind is: "Get with it or die."

In the meantime society is asking its young people to be satisfied with what they have, believe in the American Dream, and accept the heritage of genocide and pollution with pride, patriotism, and purpose. In short, we are asked to volunteer our suicides, and to do so quietly without disturbing the peace of our retiring benefactors, the over-40 generation. America was given the greatest single miracle of natural creation ever bestowed upon any civilization, but the gallery of "Great Americans" who so utterly and systematically destroyed it is a morgue celebrating the perpetuation of our fantasies of greed and power.

The consequence of genocide cannot be pardoned. The participants in that genocide cannot be excused. We do not look upon industries, churches, developers, businessmen, and politicians as being necessarily bad; we simply see them as our executioners. I am not going to befriend my executioner. I am not going to dedicate my talent and intelligence to his irresponsibility. I am going to dedicate myself to the only element that predicates our survivial and the survival of our children on down to the 10^{19} power: the stable ecology of this planet. Whatever stands in front of that goal will be destroyed. If it is the church, we will shun its halls. If it is the school, we will shut it down. If it is the bulldozers of the profit-mad congolmerates, troop trains to corrupt wars, insane commercial gluttony or the logging trucks of our paper-tiger economy that need stopping, then we will stop them. We will stop the destruction of this planet even at the cost of our own futures, careers, and blood. The situation is simply like that. If you are not going to live for the earth, what are you going to live for?

As a species we continue to commute, pollute, and salute in righteous arrogance the despoiled flag of our environment. This cannot and will not be tolerated any longer. The irony, and I hope it never becomes the tragic kind, is that never before has mankind had the tools for self-perception and global understanding that are available to it today. This statement does not, however, place the argument in the hands of the technocrats of the space-race, the bomb-now-and-study-later school of scientific panaceas, for this is surely a pitiful travesty on the true role of science in the play called "Mankind." Rather, science has given us an understanding of the evolutionary play in the ecological theater and has awakened us to a true and challenging comprehension of man and of man's place on this planet. The future, in spite of its grim portent, is the greatest hope and the greatest challenge any life form has ever had. Let it be clear, though, that the great blight of human overpopulation is the problem of success, and let us further beware lest our epitaph read: *Here lies a species that failed only because it succeeded too well.*

The misapprehension of the motives and intentions of today's young activists comes from a larger misapprehension of the age in which we live. The inner yearnings of nearly all young people are for a simple and enriching life. Coupled with the problem of global survival is the much more personal crisis of emotional survival. The cities stink. The rivers are polluted. There is no way to make an honest buck. The goal of most young people is self-realization; riddance from neuroses, anxieties, and guilt. In short, people are seeking and expressing their freedom. It is the crowning achievement of democratic culture; it is for the most part a tremendously healthy thing. The unhealthy things are catchwords in this era: alienated, freaked-out, hung-up, and others, and take their significance with respect to which ever side of the "gap" you happen to be on.

The second part of this urge to emotional wholeness and survival takes the form of a large-scale exodus from the cities to the country, but this cannot last either: there simply isn't enough country. The consequences of this step-by-step introduction to the spiritual, emotional, and physical nourishment of the undeveloped, ecologically whole countryside will be an ever greater demand for access to our natural areas, for more natural areas, and for the information, sustenance, and peace they provide. The ecological perspective provides a picture of life that focuses on a miracle of creation and evolution that is wonderful, brutal, and inspiring.

Where, one may ask, is the activism of youth heading? It is certain that the ecological perspective and the reality of the ecological crisis will mature the destructive and volatile naïveté of the young leftist. The "hashish dreams of guerrilla warfare" based on lineal Marxist pollutionist dogma are a tunnel vision to a sign reading "no exit." The real revolution is the one already under way on global food chains and on our as yet unborn.

The constuctive nature of student involvement with the issues stemming from environmental awareness is emphasized in the demands of the following manifesto composed by the youth delegates to a recent conference.

> "On a national scale, we urge:
> "The mobilization of the national effort to attain stability of numbers, and equilibrium between man and nature, by a specified date, with the attainment of this goal to be the guide for local and national policy in the intervening years;
> "The immediate assumption of a massive, federally financed study to determine the optimum carrying capacity of our country, on the community, city, county, state, and national levels, with this carrying capacity to be predicated on the quality of life, the impact upon world resources, and the tolerance of natural systems;
> "The adoption of new measures of national well-being, incorporating indices other than the rate of growth of the gross national product, the consumption of energy resources, and international credit ratings;
> "The immediate rejection of international economic competition as valid grounds for the creation of national policy.

"On an international scale, we endorse:

"The proposal that the leaders of all nations through the United Nations General Assembly declare that a state of environmental emergency exists on the planet earth;

"The creation of colleges of human ecology and survival sciences in the member nations of the United Nations;

"The creation of national, regional, and global plans for the determination of optimum population levels and distribution patterns;

"The creation of national, regional, and worldwide commissions on environmental deterioration and rehabilitation;

"The proposal that the United Nations General Assembly adopt a convenant of ecological rights similar to the U.N. covenant of human rights."

Within the changing fabric of activism itself, there is a great role yet to be played by the conservationists. It is to these people that the maturing young are going to look for help, education, and leadership. It is truly to "the men of the earth," to the men of global understanding and international commitment, that the reins of world leadership will be handed. This is the one area where the cooperation of all sides can be gained and the only area where the power structure can communicate and join forces with today's enthusiastic young activists. Without this coming together over the common goals of a quality environment and a stable ecology, we will continue to suffer the ravages of confrontation and disruption only to reap the grim harvest of irredeemable waste of energy, intelligence, and human life.

This "Student Manifesto on the Environment" is a warning . . . but more than that, it is a supplication.

The recent enthusiasm which many Americans have developed for conservation and for the preservation of the global environment is quite impressive. Yet few in America—or anywhere else—seem eager to give up their cars, their TV sets, or the rest of the flood of goods and services made possible by advanced technology. Furthermore, it is frequently pointed out that in our era all nations have dedicated themselves to the same goal: the development of an industrial society. If this be true, and if each country must have its own steel mills, its own electrical power grid, and its own jet-equipped airlines, then it appears likely that during the next few decades the earth's irreplaceable resources will be squandered and its oceans and atmosphere will be irretrievably polluted. Thus technology's ultimate gift to mankind might be a variety of methods for rendering the planet unfit for habitation, human or otherwise. Lord Ritchie-Calder, an expert on international relations and a frequent participant in United Nations activities concerning scientific and environmental affairs, analyses this threatening situation.

Lord Ritchie-Calder on Mortgaging the Old Homestead●

Past civilizations are buried in the graveyards of their own mistakes, but as each died of its greed, its carelessness or its effeteness another took its place. That was because such civilizations took their character from a locality or region. Today ours is a global civilization; it is not bounded by the Tigris and the Euphrates nor even the Hellespont and the Indus; it is the whole world. Its planet has shrunk to a neighborhood round which a man-made satellite can patrol sixteen times a day, riding the gravitational fences of Man's family estate. It is a community so interdependent that our mistakes are exaggerated on a world scale.

For the first time in history, Man has the power of veto over the evolution of his own species through a nuclear holocaust. The overkill is enough to wipe out every man, woman and child on earth, together with our fellow lodgers, the

●Source: Lord Ritchie-Calder, "Mortgaging the Old Homestead," *Foreign Affairs*, January, 1970 pp. 207-220. Reprinted by permission. Copyright © The Council on Foreign Relations, Inc. New York.

animals, the birds and the insects, and to reduce our planet to a radioactive wilderness. Or the Doomsday Machine could be replaced by the Doomsday Bug. By gene-manipulation and man-made mutations, it is possible to produce, or generate, a disease against which there would be no natural immunity; by "generate" is meant that even if the perpetrators inoculated themselves protectively, the disease in spreading round the world could assume a virulence of its own and involve them too. When a British bacteriologist died of the bug he had invented, a distinguished scientist said, "Thank God he didn't sneeze; he could have started a pandemic against which there would have been no immunity."

Modern Man can outboast the Ancients, who in the arrogance of their material achievements built pyramids as the gravestones of their civilizations. We can blast our pyramids into space to orbit through all eternity round a planet which perished by our neglect.

A hundred years ago Claude Bernard, the famous French physiologist, enjoined his colleagues, "True science teaches us to doubt and in ignorance to refrain." What he meant was that the scientist must proceed from one tested foothold to the next (like going into a mine-field with a mine-detector). Today we are using the biosphere, the living space, as an experimental laboratory. When the mad scientist of fiction blows himself and his laboratory sky-high, that is all right, but when scientists and decision-makers act out of ignorance and pretend that it is knowledge, they are putting the whole world in hazard. Anyway, science at best is not wisdom; it is knowledge, while wisdom is knowledge tempered with judgment. Because of overspecialization, most scientists are disabled from exercising judgments beyond their own sphere.

A classic example was the atomic bomb. It was the Physicists' Bomb. When the device exploded at Alamogordo on July 16, 1945, and made a notch-mark in history from which Man's future would be dated, the safe-breakers had cracked the lock of the nucleus before the locksmiths knew how it worked. (The evidence of this is the billions of dollars which have been spent since 1945 on gargantuan machines to study the fundamental particles, the components of the nucleus; and they still do not know how they interrelate.)

Prime Minister Clement Attlee, who concurred with President Truman's decision to drop the bomb on Hiroshima, later said: We knew nothing whatever at that time about the genetic effects of an atomic explosion. I knew nothing about fall-out and all the rest of what emerged after Hiroshima. As far as I know, President Truman and Winston Churchill knew nothing of those things either, nor did Sir John Anderson who coördinated research on our side. Whether the scientists directly concerned knew or guessed, I do not know. But if they did, then so far as I am aware, they said nothing of it to those who had to make the decision."[1]

---

[1] "Twilight of Empire," by Clement Attlee with Francis Williams. New York: Barnes, 1961, p. 74.

That sounds absurd, since as long before as 1927, H. J. Muller had been awarded the Nobel Prize for his evidence of the genetic effects of radiation.* But it is true that in the whole documentation of the British effort, before it merged in the Manhattan Project, there is only one reference to genetic effects—a Medical Research Council minute which was not connected with the bomb they were intending to make; it concerned the possibility that the Germans might, short of the bomb, produce radioactive isotopes as a form of biological warfare. In the Franck Report, the most statesmanlike document ever produced by scientists, with its percipience of the military and political consequences of unilateral use of the bomb (presented to Secretary of State Henry L. Stimson even before the test bomb exploded), no reference is made to the biological effects, although one would have supposed that to have been a very powerful argument. The explanation, of course, was that is was the Physicists' Bomb and military security restricted information and discussion to the bomb-makers, which excluded the biologists.

The same kind of breakdown in interdisciplinary consultation was manifest in the subsequent testing of fission and fusion bombs. Categorical assurances were given that the fallout would be confined to the testing area, but the Japanese fishing-boat *Lucky Dragon* was "dusted" well outside the predicted range. Then we got the story of radiostrontium. Radiostrontium is an analogue of calcium. Therefore in bone-formation an atom of natural strontium can take the place of calcium and the radioactive version can do likewise. Radiostrontium did not exist in the world before 1945; it is a man-made element. Today every young person, anywhere in the world, whose bones were forming during the massive bomb-testing in the atmosphere, carries this brandmark of the Atomic Age. The radiostrontium in their bones is medically insignificant, but, if the test ban (belated recognition) had not prevented the escalation of atmospheric testing, it might not have been.

Every young person everywhere was affected, and why? Because those responsible for H-bomb testing miscalculated. They assumed that the upthrust of the H-bomb would punch a hole in the stratosphere and that the gaseous radioactivity would dissipate itself. One of those gases was radioactive krypton, which quickly decays into radiostrontium, which is a particulate. The technicians had been wrongly briefed about the nature of the troposphere, the climatic ceiling which would, they maintained, prevent the fall-back. But between the equatorial troposphere and the polar troposphere there is a gap, and the radiostrontium came back through this fanlight into the climatic jetstreams. It was swept all round the world to come to earth as radioactive rain, to be deposited on foodcrops and pastures, to be ingested by animals and to get into milk and into babies and children and adolescents whose growing bones were hungry for calcium or its equivalent strontium, in this case radioactive.

---

*Editor's note: While Muller studied genetics and radiation in 1927, he was not awarded the Nobel Prize until 1946.

Incidentally, radiostrontium was known to the biologists before it "hit it headlines." They had found it in the skin burns of animals exposed on the Nevada testing ranges and they knew its sinister nature as a "boneseeker." But the authorities clapped security on their work, classified it as "Operation Sunshine" and cynically called the units of radiostrontium "Sunshine Units"—an instance not of ignorance but of deliberate non-communication.

One beneficial effect of the alarm caused by all this has been that the atoms industry is, bar none, the safest in the world for those working in it. Precautions, now universal, were built into the code of practice from the beginning. Indeed it can be admitted that the safety margins in health and in working conditions are perhaps excessive in the light of experience, but no one would dare to modify them. There can, however, be accidents in which the public assumes the risk. At Windscale, the British atomic center in Cumberland, a reactor burned out. Radioactive fumes escaped from the stacks in spite of the filters. They drifted over the country. Milk was dumped into the sea because radioactive iodine had covered the dairy pastures.

There is the problem of atomic waste disposal, which persists in the peaceful uses as well as in the making of nuclear explosives. Low energy wastes, carefully monitored, can be safely disposed of. Trash, irradiated metals and laboratory waste can be embedded in concrete and dumped in the ocean deeps—although this practice raises more misgivings. But high-level wastes, some with elements the radioactivity of which can persist for *hundreds of thousands* of years, present prodigious difficulties. There must be "burial grounds" (or euphemistically "farms"), the biggest of which is at Hanford, Washington. It encloses a stretch of the Columbia River in a tract covering 650 square miles, where no one is allowed to live or to trespass.

There, in the twentieth century Giza, it has cost more, much more, to bury live atoms than it cost to entomb the sun-god Kings of Egypt. The capital outlay runs into hundreds of millions of dollars and the maintenance of the U.S. sepulchres is over $6 million a year. (Add to that the buried waste of the U.S.S.R., Britain, Canada, France, and China, and one can see what it costs to bury live atoms.) And they are very much alive. At Hanford they are kept in million-gallon carbon-steel tanks. Their radioactive vitality keeps the accompanying acids boiling like a witches' cauldron. A cooling system has to be maintained continuously. The vapors from the self-boiling tanks have to be condensed and "scrubbed" (radioactive atoms removed); otherwise a radioactive miasma would escape from the vents. The tanks will not endure as long as the pyramids and certainly not for the hundreds of thousands of years of the longlived atoms. The acids and the atomic ferments erode the toughest metal, so the tanks have to be periodically decanted. Another method is to entomb them in disused salt mines. Another is to embed them in ceramics, lock them up in glass beads. Another is what is known as "hydraulic fraction": a hole is drilled into a shale formation (below the subsoil water); liquid is piped down under pressure and causes the shale to split laterally. Hence the atoms in liquid cement can be injected under enormous pressure and spread into the fissures to set like a radioactive sandwich. This overcomes the worry of earthquakes since the materials will remain secure even in faulting.

This accumulating waste from fission plants will persist until the promise, still far from fulfilled, of peaceful thermonuclear power comes about. With the multiplication of power reactors, the wastes will increase. It is calculated that by the year 2000, the number of six-ton nuclear "hearses" in transit to "burial grounds" at any given time on the highways of the United States will be well over 3,000 and the amount of radioactive products will be about a billion curies, which is a mighty lot of curies to be roaming around a populated country.

The alarming possibilities were well illustrated by the incident at Palomares, on the coast of Spain, when there occurred a collision of a refueling aircraft with a U. S. nuclear bomber on "live" mission. The bombs were scattered. There was no explosion, but radioactive materials broke loose and the contaminated beaches and farm soil had to be scooped up and taken to the United States for burial.

Imagine what would have happened if the *Torrey Canyon,* the giant tanker which was wrecked off the Scilly Isles, had been nuclear-powered. Some experts make comforting noises and say that the reactors would have "closed down," but the *Torrey Canyon* was a wreck and the Palomares incident showed what happens when radioactive materials break loose. All those oil-polluted beaches of southwest England and the coasts of Brittany would have had to be scooped up for nuclear burial.

II

The *Torrey Canyon* is a nightmarish example of progress for its own sake. The bigger the tanker the cheaper the freightage, which is supposed to be progress. This ship was built at Newport, Virginia, in 1959 for the Union Oil Corporation; it was a giant for the time—810 feet long and 104 feet beam—but five years later, that was not big enough. She was taken to Japan to be "stretched." The ship was cut in half amidship and a mid-body section inserted. With a new bow, this made her 954 feet long, and her beam was extended 21 feet. She could carry 850,000 barrels of oil, twice her original capacity.

Built for Union Oil, she was "owned" by the Barracuda Tanker Corporation, the head office of which is a filing cabinet in Hamilton, Bermuda. She was registered under the Liberian flag of convenience and her captain and crew were Italians, recruited in Genoa. Just to complicate the international tangle, she was under charter to the British Petroleum Tanker Company to bring 118,000 tons of crude oil from Kuwait to Milford Haven in Wales, via the Cape of Good Hope.

Approaching Lands End, the Italian captain was informed that if he did not reach Milford Haven by 11 p.m. Saturday night, he would miss high-water and would not be able to enter the harbor for another five days, which would have annoyed his employers. He took a shortcut, setting course between Seven Stones rocks and the Scilly Isles, and he finished up on Pollard Rock, in an area where no ship of that size should ever have been.

Her ruptured tanks began to vomit oil and great slicks spread over the sea in the direction of the Cornish holiday beaches. A Dutch tug made a dash for the stranded ship, gambling on the salvage money. (Where the salvaged ship could have been taken one cannot imagine, since no place would offer harborage to a leaking tanker). After delays and a death in the futile salvage effort, the British Government moved in with the navy, the air force and, on the beaches, the army. They tried to set fire to the floating oil which, of course, would not volatilize. They covered the slicks with detergents (supplied at a price by the oil companies), and then the bombers moved in to try to cut open the deck and, with incendiaries, to set fire to the remaining oil in the tanks. Finally the ship foundered and divers confirmed that the oil had been effectively consumed.

Nevertheless the result was havoc. All measures had had to be improvised. Twelve thousand tons of detergent went into the sea. Later marine biologists found that the cure had been worse than the complaint. The oil was disastrous for seabirds, but marine organic life was destroyed by the detergents. By arduous physical efforts, with bulldozers and flame-throwers and, again, more detergents, the beaches were cleaned up for the holiday-makers. Northerly winds swept the oil slicks down Channel to the French coast with even more serious consequences, particularly to the valuable shellfish industry. With even bigger tankers being launched, this affair is a portentous warning.

Two years after *Torrey Canyon* an offshore oil rig erupted in the Santa Barbara Channel. The disaster to wildlife in this area, which has island nature reserves and is on the migratory route of whales, seals and seabirds, was a repetition of the *Torrey Canyon* oil-spill. And the owner of the lethal oil rig was Union Oil.

III

Another piece of stupidity shows how much we are at the mercy of ignorant men pretending to be knowledgeable. During the International Geophysical Year, 1957-58, the Van Allen Belt was discovered. This is an area of magnetic phenomena. Immediately it was decided to explode a hydrogen bomb in the Belt to see whether an artificial aurora could be produced. The colorful draperies and luminous skirts of the aurora borealis are caused by the drawing in of cosmic particles through the rare gases of the upper atmosphere—ionization it is called; it is like passing electrons through the vacuum tubes of our familiar fluorescent lighting. The name Rainbow Bomb was given it in anticipation of the display it was expected to produce. Every eminent scientist in the field of cosmology, radio-astronomy or physics of the atmosphere protested at this irresponsible tampering with a system which we did not understand. And typical of the casual attitude toward this kind of thing, the Prime Minister of the day, answering protests in the House of Commons that called on him to intervene with the Americans, asked what all the fuss was about. After all, they hadn't known that the Van Allen Belt even existed a year before. This was the cosmic equivalent of Chamberlain's remark about Czechoslovakia, at the time of Munich, about that distant country of which we knew so little. They exploded the bomb. They got their pyrotechnics and we still do not know the cost we may have to pay for this artificial magnetic disturbance.

In the same way we can look with migivings on those tracks—the white tails of the jets, which are introducing into our climatic system new factors, the effects of which are immensurable. Formation of rain clouds depends on water vapor having a nucleum on which to form. That is how artificial precipitation is introduced—the so-called rain-making. So the jets, criss-crossing the weather system, playing noughts and crosses with it, can produce a man-made change.

In the longer term we can foresee even more drastic effects from Man's unthinking operations. At the United Nations' Science and Technology Conference in Geneva in 1963 we took stock of the effects of industrialization on our total environment thus far. The atmosphere is not only the air which humans, animals and plants breathe; it is also the envelope which protects living things from harmful radiation from the sun and outer space. It is also the medium of climate, the winds and the rain. Those are inseparable from the hydrosphere—the oceans, covering seven-tenths of the globe, with their currents and extraordinary rates of evaporation; the biosphere, with its trees and their transpiration; and in terms of human activities, the minerals mined from the lithosphere, the rock crust. Millions of years ago the sun encouraged the growth of the primeval forests, which became our coal, and the plant growth of the seas, which became our oil. Those fossil fuels, locked away for aeons of time, are extracted by man and put back into the atmosphere from the chimney stacks and the exhaust pipes of modern enginering. About 6 billion tons of carbon are mixed with the atmosphere annually. During the past century, in the process of industrialization, with its release of carbon by the burning of fossil fuels, more than 400 billion tons of carbon have been artificially introduced into the atmosphere. The concentration in the air we breathe has been increased by approximately 10 percent, and if all the known reserves of coal and oil were burnt, the concentration would be ten times greater.

This is something more than a public health problem, more than a question of what goes into the lungs of an individual, more than a question of smog. The carbon cycle in nature is a self-adjusting mechanism. Carbon dioxide is, of course, indispensable for plants and is, therefore, a source of life, but there is a balance which is maintained by excess carbon being absorbed by the seas. The excess is now taxing this absorption and it can seriously disturb the heat balance of the earth because of what is known as the "greenhouse effect." A greenhouse lets in the sun's rays but retains the heat. Carbon dioxide, as a transparent diffusion, does likewise. It keeps the heat at the surface of the earth and in excess modifies the climate.

It has been estimated that, at the present rate of increase, the mean annual temperature all over the world might increase by 3.6 degrees centrigrade in the next forty to fifty years. The experts may argue about the time factor and even about the effects, but certain things are apparent, not only in the industrialized Northern Hemisphere but in the Southern Hemisphere also. The North-polar ice-cap is thinning and shrinking. The seas, with their blanket of carbon dioxide, are changing their temperature, with the result that marine plant life is increasing and is transpiring more carbon dioxide. As a result of the combination, fish are migrating, changing even their latitudes. On land the snow line is retreating and glaciers are melting. In Scandinavia, land which was perennially under snow and ice is thawing, and arrowheads of over 1,000 years ago, when the black soils were last exposed, have been found. The melting of sea ice will not affect the sea level, because the volume of floating ice is the same as the water it diplaces, but the melting of ice caps or glaciers, in which the water is locked up, will introduce additional water to the sea and raise the level. Rivers originating in glaciers and permanent snow fields will increase their flow; and if ice dams, such as those in the Himalayas, break, the results in flooding may be catastrophic. In this process the patterns of rainfall will change, with increased precipitation in some areas and the possiblity of aridity in now fertile regions. One would be well-advised not to take ninety-nine year leases on properties at present sea level.

IV

At that same conference, there was a sobering reminder of mistakes which can be writ large, from the very best intentions. In the Indus Valley in West Pakistan, the population is increasing at the rate of ten more mouths to be fed every five minutes. In that same five minutes in that same place, an acre of land is being lost through water-logging and salinity. This is the largest irrigated region in the world. Twenty-three million acres are artificially watered by canals. The Indus and its tributaries, the Jhelum, the Chenab, the Ravi, the Beas and the Sutlej, created the alluvial plains of the Punjab and the Sind. In the nineteenth century, the British began a big program of farm development in lands which were fertile but had low rainfall. Barrages and distribution canals were constructed. One thing which, for economy's sake, was not done was to line the canals. In the early days, this genuinely did not matter. The water was being spread from the Indus into a thirsty plain and if it soaked in so much the better. The system also depended on what is called "inland delta drainage," that is to say, the water spreads out like a delta and then drains itself back into the river. After independence, Pakistan, with external aid, started vigorously to extend the Indus irrigation. The experts all said the soil was good and would produce abundantly once it got the distributed water. There were plenty of experts, but they all overlooked one thing—the hydrological imperatives. The incline from Lahore to the Rann of Kutch—700 miles— is a foot a mile, a quite inadequate drainage gradient. So as more and more barrages and more and more lateral canals were built, the water was not draining back into the Indus. Some 40 percent of the water in the unlined canals seeped underground, and in a network of 40,000 miles of canals that is a lot of water. The result was that the watertable rose. Low-lying areas became waterlogged, drowning the roots of the crops. In other areas the water crept upwards, leaching salts which accumulated in the surface layers, poisoning the crops. At the same time the irrigation régime, which used just 1-½ inches of water a year in the fields, did not sluice out those salts but added, through evaporation, its own salts. The result was tragically spectacular. In flying over large tracts of this area one would imagine that it was an Arctic landscape because the white crust of salt glistens like snow.

The situation was deteriorating so rapidly that President Ayub appealed in person to President Kennedy, who sent out a high-powered mission which encompassed twenty disciplines. This was backed by the computers at Harvard. The answers were pretty grim. It would take twenty years and $2 billion to repair the damage—more than it cost to create the installations that did the damage. It would mean using vertical drainage to bring up the water and use it for irrigation, and also to sluice out the salt in the surface soil. If those twenty scientific disciplines had been brought together in the first instance it would not have happened.

One more instance of the far-flung consequences of men's localized mistakes: No insecticides or pesticides have ever been allowed into the continent of Antarctica. Yet they have been found in the fauna along the northern coasts. They have come almost certainly from the Northern Hemisphere, carried from the rivers of the farm-states into the currents sweeping south. In November 1969, the U. S. Government decided to "phase out" the use of DDT.

Pollution is a crime compounded of ignorance and avarice. The great achievements of *Homo Sapiens* become the disaster-ridden blunders of Un-thinking Man—poisoned rivers and dead lakes, polluted with effluents of industries which give somthing called "prosperity" at the expense of posterity. Rivers are treated like sewers and lakes like cesspools. These natural systems— and they are living systems—have struggled hard. The benevolent microorganisms which cope with reasonable amounts of organic matter have been destroyed by mineral detergents. Witness our foaming streams. Lake Erie did its best to provide the oxygen to neutralize the pickling acids of the great steel works. But it could not contend. It lost its oxygen in the battle. Its once rich commerical fishing industry died and its revitalizing microorganic life gave place to anaerobic organisms which do not need oxygen but give off foul smells, the mortuary smells of dead water. As one Erie industrialist retorted, "It's not our effluent; it's those damned dead fish."

We have had the Freedom from Hunger Campaign; presently we shall need a Freedom from Thirst Campaign. If the International Hydrological Decade does not bring us to our senses we will face a desperate situation. Of course it is bound up with the increasing population but also with the extravagances of the technologies which claim that they are serving that population. There is a competition between the water needs of the land which has to feed the increasing population and the domestic and industrial needs of that population. The theoretical minimum to sustain living standards is about 300 gallons a day per person. This is the approximate amount of water needed to produce grain for 2½ pounds of bread, but a diet of 2 pounds of bread and 1 pound of beef would require about 2,500 gallons. And that is nothing compared with the gluttonous requirements of steelmaking, paper-making and the chemical industry.

Water—just H_2O—is as indispensable as food. To die of hunger one needs more than fifteen days. To die of thirst one needs only three. Yet we are squandering, polluting and destroying water. In Los Angeles and neighboring Southern California, a thousand times more water is being consumed than is being precipitated in the locality. They have preëmpted the water of neighboring states. They are piping it from Northern California and there is a plan to pipe it all the way from Canada's Northwest Territories, from the Mackenzie and the Liard which flow northwards to Arctic Ocean, to turn them back into deserts.

V

Always and everywhere we come back to the problem of population —more people to make more mistakes, more people to be the victims of the mistakes of others, more people to suffer Hell upon Earth. It is appalling to hear people complacently talking about the population explosion as though it belonged to the future, or world hunger as though it were threatening, when hundreds of millions can testify that it is already here—swear it with panting breath.

We know to the exact countdown second when the nuclear explosion took place—5:30 a.m., July 16, 1945, when the first device went off in the desert of

Alamogordo, New Mexico. The fuse of the population explosion had been lit ten years earlier—February 1935. On that day a girl called Hildegarde was dying of generalized septicaemia. She had pricked her finger with a sewing needle and the infection had run amok. The doctors could not save her. Her desperate father injected a red dye into her body. Her father was Gerhard Domagk. The red dye was prontosil which he, a pharmaceutical chemist, had produced and had successfully used on mice lethally infected with streptococci, but never before on a human. Prontosil was the first of the sulfa drugs—chemotherapeutics, which could attack the germ within the living body. Thus was prepared the way for the rediscovery of penicillin—rediscovery because although Fleming had discovered it in 1928, it had been ignored because neither he nor anybody else had seen its supreme virtue of attacking germs within the living body. That is the operative phrase, for while medical science and the medical profession had used antiseptics for surface wounds and sores, they were always labeled "Poison, not to be taken internally." The sulfa drugs had shown that it was possible to attack specific germs within the living body and had changed this attitude. So when Chain and Florey looked again at Fleming's penicillin in 1938, they were seeing it in the light of the experience of the sulphas.

A new era of disease-fighting had begun—the sulfas, the antibiotics, DDT insecticides. Doctors could now attack a whole range of invisible enemies. They could master the old killer diseases. They proved it during the war, and when the war ended there were not only stockpiles of the drugs, there were tooled up factories to produce them. So to prevent the spread of the deadly epidemics which follow wars, the supplies were made available to the war-ravaged countries with their displaced persons, and then to the developing countries. Their indigenous infections and contagions and insect-borne diseases were checked.

Almost symbolically, the first great clinical use of prontosil had been in dealing with puerperal sepsis, childbed fever. It had spectacularly saved mothers' lives in Queen Charlotte's Hospital, London. Now its successors took up the story. Fewer mothers died in childbirth, to live and have more babies. Fewer infants died, fewer toddlers, fewer adolescents. They lived to marry and have children. Older people were not killed off by, for instance, malaria. The average life-span increased.

Professor Kingsley Davis of the University of California at Berkeley, the authority on urban development, has presented a hair-raising picture from his survey of the world's cities. He has shown that 38 percent of the world's population is already living in what are defined as urban places. Over one-fifth of the world's population is living in cities of 100,000 or more. And over one-tenth of the world's population is now living in cities of a million or more inhabitants. In 1968, 375 million people were living in million-and-over cities. The proportions are changing so quickly that on present trends it would take only 16 years for half the world's population to be living in cities and only fifty-five years for it to reach 100 percent.

Within the lifetime of a child born today, Kingsley Davis foresees, on present trends of population-increase, 15 billion people to be fed and housed—nearly five times as many as now. The whole human species would be living in cities of a million-and-over inhabitants, and—wait for it!—the biggest city would have 1.3 billion inhabitants. That means 186 times as many as there are in Greater London.

For years the Greek architect Doxiadis has been warning us about such prospects. In his Ecumenopolis—World City—one urban area like confluent ulcers would ooze into the next. The East Side of World City would have as its High Street the Eurasian Highway stretching from Glasgow to Bangkok, with the Channel Tunnel as its subway and a built-up area all the way. On the West Side of World City, divided not by the tracks but by the Atlantic, the pattern is already emerging, or rather, merging. Americans already talk about Boswash, the urban development of a built-up area stretching from Boston to Washington; and on the West Coast, apart from Los Angeles, sprawling into the desert, the realtors are already slurring one city into another all along the Pacific Coast from the Mexican Border to San Francisco. We don't need a crystal ball to foresee what Davis and Doziadis are predicting; we can already see it through smog-covered spectacles; a blind man can smell what is coming.

The danger of prediction is that experts and men of affairs are likely to plan for the predicted trends and confirm these trends. "Prognosis" is something different from "prediction." An intelligent doctor having diagnosed your sysmptoms and examined your condition does not say (except in novelettes), "You have six months to live." An intelligent doctor says, "Frankly, your condition is serious. Unless you do so-and-so, it is bound to deteriorate." The operative pharse is "do so-and-so." We don't have to plan for trends; if they are socially undesirable our duty is to plan away from them; to treat the symptoms before they become malignant.

We have to do this on the local, the national and the international scale, through intergovernmental action, because there are no frontiers in present-day pollution and destruction of the biosphere. Mankind shares a common habitat. We have mortgaged the old homestead and nature is liable to foreclose.

"Behold now Behemoth . . ."

Book of Job

"The horrifying truth is that, so far as much technology is concerned, no one is in charge."

Alvin Toffler●

VI. A Question of Control

To some observers, the technological revolution has had unfortunate effects going far beyond pollution and resource depletion. Accelerating social change can be held responsible for the crisis of the cities, juvenile delinquency, racial problems, and even the chaotic state of arts and letters. This is basically the position of Mr. Michael Harrington, an American socialist who is frequently credited with having rediscovered and publicized the continued existence of poverty within our affluent economy. He argues that the crucial problems afflicting our society result from drift and purposelessness. Runaway technology has been allowed to charge onward, unguided by rational debate or democratic choice, and it has created, willy-nilly, an "accidental revolution." Thus, having muddled along without direction or decision, we find ourselves confronted with a multitude of grave social ills. But Mr. Harrington has faith that man *can* take control of the system, that technology can be subjected to democratic control and directed to humane ends, and that ". . . the future would thus be chosen rather than submitted to."

●Source: Reprinted by permission of Random House, Inc., from *Future Shock* by Alvin Toffler. Copyright © by Alvin Toffler, 1970.

Michael Harrington on
the Accidental Revolution*

* * * * *

This accidental revolution is the sweeping and unprecedented technological transformation of the Western environment which has been, and is being, carried out in a casual way. In it, this technology is essentially under private control and used for private purposes; this situation is justified in the name of a conservative ideology; and the by-product is a historical change which would have staggered the imagination of any nineteenth-century visionary. In following their individual aims, industrialists blundered into a social revolution. There is indeed an invisible hand in all of this. Only it is shaping an unstable new world rather than Adam Smith's middle-class harmony.

But this comforting analogy misses the very essence of the accidental revolution. The older ideologists and utopians were victimized by their ignorance of the limits of the possible. They sought divine commonwealths and secular salvations which were impossible of achievement. Where these conscious revolutionists of the past proposed visions which outstripped reality, the unconscious revolutionists of the present create realities which outstrip their vision. In the first case, it is history that is sad, in the second, man.

As a result of this development, cracks opened up in every ideology and philosophy. Conservatives unwittingly made a revolution, but it was not the one the revolutionists had predicted, and the antagonists were mutually bewildered. Religious thinkers reported the progress of godlessness, but the fact did not seem heroic to atheists. Literacy increased, making many educators fearful, since its uses seemed anticultural; the people asserted themselves and the traditional democrats became uneasy. There was perplexity on all sides. And many decadences were thus discovered.

What is in decay in these theories is not so much the past as the future. To be sure, some conservatives nostalgically looked back to the golden days of happy hierarchies that never really existed. But, as shall be seen, the more serious idea of decadence is that the West no longer senses either a City of God or of man in the middle or long distance. It has lost its utopia to come rather than its golden

•Source: Reprinted with permission of The Macmillan Company from *The Accidental Century* by Michael Harrington. Copyright © by Michael Harrington, 1965 pp. 16-42. (Sections have been omitted with permission of the Publisher.)

age that was. And this is the meaning of the term intended here: the present decadence is the corruption of a dream rather than of a reality.

The city is a prosaic, and crucial, case in point of the accidental revolution. To the Germans it was the very climax of decline. More broadly, it is one of the most accessible examples of the contemporary predicament. In what follows, a fairly familiar description is summarized as an illustration of the not-so-obvious workings of the accidental revolution.

In the last decade or so, social scientists have invented a word to describe a new stage of urban life: Megalopolis. In part, this term reflects the sheer and tumultuous growth in the population of cities. In 1800, there did not exist a city of a million inhabitants, and London, with 959,310 citizens, was the largest concentration of people on the face of the earth. By 1850, London numbered two million; by 1910, there were eleven places in the world with over a million inhabitants; by 1930, twenty-seven which had exceeded that limit. And the trend, of course, continues.

In and of themselves, these quantities gave rise to the Malthusian theory of decadence. Population, it was said, would outstrip food supply; the birthrate was a harbinger of social death. If these fears have diminished in the advanced nations of the West, which are the subject of this book, they have become more plausible on a global scale. In the characteristic mode of the accidental revolution, science makes more existences possible, but casually, without bothering with the consequences.

But the main thrust of the concept of Megalopolis in the advanced lands does not derive from the Malthusian fear. It defines a new system of Western social relations. As the cities grew, more and more people—the fortunate ones—left the central city and went to the suburbs. Urban life expanded from the old industrial center, leaving decaying areas in its wake, and migrated to the new, low-population fringes. Eventually, even the factories joined in the pattern, and a vast system of highways supported two huge streams of daily traffic, the blue-collar out-commuters hurrying to the margins, the executive and white-collar mass speeding to the downtown business area. Because of the vast, sprawling spaces involved, mass transportation became more costly and inefficient, and a supercongestion became a normal feature of life. (Throughout this book, the preponderance of evidence will be American, but that is only because the United States, as the most technologically advanced nation, shows the farthest flung trends of advanced society in general.)

These events also transformed the landscape. In the United States, Megalopolis defined urban continuums hundreds of miles long (for instance, from Boston in the North to Washington, D. C. in the South). The city, like man, no longer had limits. It ceased to be a nucleus of civilization set in the countryside and reached out, obliterating the immemorial distinction between town and nature. As a result, man could not escape himself.

All this had profound consequences for every aspect of life. It tends, as I tried to show in my book, *The Other America,* to alter the very eyes of society. Those left behind in the central city—the aged, the racial and national minorities, the poor generally—drop out of the mind and sight of those riding the super-highways from suburbia to office and back. On a broader, economic level, entire regions were left, like neighborhoods, to rot and stagnate in the midst of rapid change: the American Appalachians, the English North, the Italian South. Intimately, Megalopolis, and particularly its automobile, helped to rearrange the structure of family life and played a major role in the destruction of the traditional sexual ethic.

The nineteenth-century seers had argued that such fundamental changes in the human environment would create different kinds of people. This happened. But, typically, the new personality types did not correspond to the old hopes any more than the new society did. There is, for example, that most unlikely revolutionary creation, the teen-ager. Teen-age is not a chronological phenomenon—there have always been young people between twelve and twenty—but a historical one, the product of the accidental revolution. As a result of Megalopolis and the technological culture which it incarnated, there appeared a generation of adolescents with leisure, money, and mobility. They constituted a huge market, and tastes were duly fabricated for them. And they had a radical impact on the very quality of social life.

America was, of course, the land of the teen-age pioneers. But as American standards of living began to spread throughout the West, so did the styles of the American young. The teen-ager, an afterthought of social history, became a more genuine example of the "new man" than any of the calculated personalities which the Communists held up for emulation. The Russians discovered that even totalitarianism could not keep out jazz and cool music. And there were other kinds of people invented in the United States and exported to Europe: the organization man, the suburban housewife, and so on.

One of the most important aspects of this chaotic emergence of un-precedented ways of living and artificial countrysides was that it was governmentally encouraged but not democratically planned. In the United States in the postwar period, huge Federal subsidies were granted to the suburban rush of the middle class and the rich. An express Congressional commitment to a broad, low-cost housing program for the poor was made in 1949 and then usually ignored. In a good many instances, the impoverished slum-dwellers, and the Negroes above all, were the prime victims of an urban renewal which was supposed to help them. In England during the fifties, there was a parallel policy of official support for office buildings and automobile congestion and against homes and quiet. By the early sixties in Paris, the telltale Megalopolitan apartments looked down on the gray, working-class *banlieues.* (Marc Paillet, in his 1964 study, *Gauche, Année Zéro,* estimated the annual French deficit in new housing units at 150,000 out of 500,000 required.)

It took the power of society to facilitate changes of such a magnitude. Yet, even though the collectivity was deeply involved in the causes and effects of Megalopolis, there was never a free, thorough debate on the transformation. The priorities of the future were derived from the *status quo,* which satisfied neither the conservatives nor the revolutionists.

The resulting continuity of Megalopolitan problems within the West provoked like-sounding indictments. In America, Lewis Mumford wrote of "the increasing pathology of the whole mode of life in the great metropolis, a pathology that is directly proportionate to its overgrowth, its purposeless materialism, its congestion and insensate disorder . . . That sinister state manifests itself not merely in the statistics of crime and mental disorder, but in the enormous sums spent on narcotics, sedatives, stimulants, hypnotics and tranquilizers to keep the population of our 'great cities' from coming to terms with the vacuous desperation of their daily lives and the even more vacuous horrors that their rulers and scientific advisers seem to regard as a reasonable terminal for the human race."

And the British social critic, C. A. R. Crosland, said, "Greedy men, abetted by a complaisant government, are prowling over Britain and devastating it . . . Excited by speculative gain, the property developers furiously rebuild the urban centers with unplanned and tawdry office-blocks; so our cities became the just objects of world-wide pity and ridicule for their architectural mediocrity, commercial vulgarity, and lack of civic or historic pride."

The coming of Megalopolis wrote a cruel denouement to many a nineteenth-century vision. For what happened was that a revolution took place without conscious revolutionists.

Megalopolis and the age it represented constituted a most radical re-structuring of the experience of life itself. There has truly been a "devaluing of all values" as Nietzsche said there would be. But the event did not summon up a new race of stoic and aristocratic supermen. The barbarians who acted in Nietzsche's name would have been despised by their master. Capitalism had, as Marx announced, been unable to contain its own technology within the bounds of its traditional theory and practice, but it had not been succeeded by a socialist leap into freedom. In short, the material transformations have exceeded the wildest imaginings of the science and social fiction of the last century—and so has the conservatism of men. Instead of emancipated proletarians, as in Marx, or sensitive slaveowners as in Nietzsche, there are teen-agers, organization men, suburban housewives.

How had this come to pass? Through technology, some answered, mistaking a precondition for a cause.

It is a cliche that the West (and throughout, I will take the term to mean Europe and America) has been undergoing a continuous technological revolution in this century. In Henry Adams' curious, but revealing, metaphor, time is speeding up, history is becoming volatile, like water in the transition from ice to liquid to steam. Around 1900, Adams calculated that the next, and decisive, phase in the process would take place in the middle of 1917, which was a remarkable intuition no matter how eccentric his method. But, whatever point of departure one chooses, it is clear that during the last sixty or so years Western man has been refashioning reality and, often without noticing it, himself.

And technology was, of course, a basic element in this revolution. First the symbol was the railroad. "We who lived before the railway," Thackeray wrote, "and survive out of the ancient world, are like Father Noah and his family out of the Ark. The children will gather around and say to us patriarchs, 'Tell us, Grandpa, about the old world.'" Then came the automobile, both as a mighty achievement of mass production and as a social force in its own dynamic right. And then the airplane and then the rocket and no one knows what next. There was a succession of technological floods, and each new generation looked back to its youth like Father Noah.

As John Maynard Keynes put the radical character of this development: "At some epoch before the dawn of history—perhaps even in one of the comfortable intervals before the last ice age—there must have been an era of progress and invention comparable to those in which we live today." Between that hypothesized burst of primitive ingenuity and the eighteenth century, Keynes said, "there was no very great change in the standard of life of the average man in the civilized centers of the earth." Then, with gathering momentum, man in the last two centuries changed himself more basically than at any time since the beginning of recorded history. To cite a typical prodigy, in Colin Clark's figures, net income from manufacturing in the United States rose by 4,500 percent between 1860 and 1953.

And all the signs point, of course, to an acceleration of this process in the future. In 1960 a group of scientists (Nobel Prize winners, members of the Moscow Academy of Science, and others) predicted the state of invention in the year 2000. Voyages to the moon, they said, would be normal, and there would be inhabited artificial satellites. The population of the world would have increased fourfold and stabilized. Seawater and rocks will yield all necessary metals, and knowledge will be accumulated in electronic banks and transmitted directly to the human nervous system by means of coded electronic messages. And genetics will allow the scientific planning of the personalities of the next generation.

There is no need to prolong the recital. The miraculousness of technology is a popular article of Western faith. Henry Adams long ago realized that the modern dynamo was an alternate principle of civilization to the medieval Virgin—only he did not understand how soon the dynamo would become old-fashioned. And yet, it is not machines, but the uses dictated to them by men which create new societies. The responsibility for the accidental revolution is human, not inanimate.

In order to understand this fact, it is necessary to speak of capitalism. This will come as an embarrassment to many in the West, and particularly in the United States, for there is a strong tendency to doubt or ignore the existence of such a system. (Some of the reasons for this attitude will be discussed shortly.) To others, the introduction of the term is an announcement that caricature will follow, another one of those dreary, monomaniacal reductions of the complexity of life to statistics about production.

A few things should be made clear. The private despairs, the aesthetic creations, the personal psychologies of these times are not simply the reflection of an economic system. They are, indeed, not "simply" any one thing, for they are always complicated, and they often have their own autonomies and cannot be understood without being taken seriously on their own terms. At the same time, these arts and inner selves do not take place in a void but in a given century and in a century that would not let any man alone. The various aspects of decadence cannot be turned into functions of the economy and neither can they be grasped without relating them to this dynamic and momentous part of contemporary life.

To talk of capitalism as an economic system is the first step away from fatalism. If it is the machines alone which have created all these changes, then the best that one can do is to pray to the computers and production lines that they will become more benign. Such an approach leads to a modern animism that invests technology with the spirits that once inhabited trees and storms. But if it is man's use of machines—the economic system—that is responsibile for what is happening, then the direction of events can be altered. And that is a basic premise of this optimistic study of decadence.

There is no question that Western capitalism has changed enormously from the nineteenth-century model. Some of these transformations are analyzed in Chapter 3. Child labor has long been legally abolished, most of the sweatshops are gone, there has been a persistent decline in the length of the working day, and so on. In addition, the night-watchman state of the *laissez-faire* ideal has given way, in one degree or another, to the welfare state in every advanced nation. In the place of the robber baron and the plutocrat—"I owe the public nothing," J.P. Morgan said—there is the corporate technocrat giving speeches on the social responsibility of his firm and the partnership of capital and labor, whether hypocritically or not.

Yet, however genteel these modifications have made the system, the basic allocation of resources is still made in the pursuit of profit. Production decisions are reckoned in terms of private advantage without reference to social consequences, at least so far as that is politically possible. There is a considerable literature which examines this calculus of gain and how it has been altered by the separation of ownership and management in the corporate collective. But, whatever theory one chooses, the corporation has not become philanthropic or democratic. In one way or another, the executive, no matter how responsible, is paid for making money and not for guiding technological change in a humane and decent way.

Without getting into unnecessary arguments, the force of this aspect of the accidental revolution can be put in terms of a familiar distinction. It has long been recognized that under capitalism there can be a divergence between the private and social cost of a good or service. In the dear, dead old days, this divorce appeared manageable and measurable. A railroad, in its pursuit of profit, would start a new line. It would cost so much money for the materials, the engineering, and the land. But then, there would be another cost. The sparks from the train would injure trees bordering the track. Who was to pay, the company or the farmer? That was a relatively easy issue, and it could be settled in the courts. The new line would destroy the beauty of a valley and deprive an entire town of its view. That was more distant, less amenable to the assignment of responsibility.

But now, as the twentieth century advances, a chasm opens up between private and social cost. The production of automobiles play a role in changing the structure of the family, sexual mores, and polluting the very air men breathe. The private decision of real-estate developers to ring a city with carefully zoned, relatively expensive suburbs exacerbates racial tensions, segregates education on the basis of color and class, modifies the urban tax base and consequently the political order, and embitters the experience of old age for those who are left behind.

Take an example of this same phenomenon involving an entire geographic region. In Appalachia, the post-World War II decision of the coal companies to automate had the most profound consequences. The new technology of strip and bore mining literally tore up the very landscape, decreased mining employment by almost two-thirds, made various forms of traditional agriculture impossible, eroded the land-tax basis of the already inadequate school system, drove hundreds of thousands of people into cities and an urban life for which they were totally unprepared, and even dangerously changed the very quality of the water that flowed down into the reservoirs of the Tennessee Valley. But the courts held that corporations had the private right to work this public havoc because of contracts that had been signed at least a generation before anyone had the remotest idea of what they would come to mean. (All this, and much more, is documented in Harry Caudil's brilliant book, *Night Comes to the Cumberlands.*)

This is one way of saying that the contemporary technology is, of its very radical nature, social. If it is left within a context of private, often hit-or-miss, decision-making, that is not so much free enterprise as it is the rule of corporate bureaucracy in the public sphere. And it is precisely this mechanism which gives the present revolution its accidental quality. The cause of change—a personal or corporate investment in gain, a private cost and profit—stands in little or no relation to the effect—a new order of human life.

One of the most glaring examples of this reality has already been cited: the heedless speculation in land and building which has characterized practically every postwar Western nation. (An exception, like Sweden, receives its privileged position precisely from the conscious intrusion of a public motive in this area and from a large, cooperative housing sector.) And so the desperate needs of slum-dwellers are sacrificed to the profitable luxuries of middle- and upper-income homes.

In the market sector, there is a gigantic apparatus designed to mask this corporate egotism. Through one of society's most important educators, the advertising industry, all of the techniques of science are used for the private socialization of the public taste. The consumer's "free" choice is thus engineered and calculated as far as is possible so that it will coincide with the highest profitability to the producer. The influence of this carefully wrought value system increases as one descends the social class ladder. As in all things, the poor pay the highest cost and give aid and comfort to the rich.

In any case, the growing chasm between private and social cost, the corporate control of a technology which is essentially public, called into question a most basic assumption of Western thought: that man frees himself through reason. Each new invention, like the automobile, television, nuclear power, space rockets, was a triumph of the human intelligence. But the totality of these innovations, with all of their revolutionary consequences, was an increasingly puzzling, even mysterious, society. As the parts became more ingenious and minutely calibrated, the whole became more irrational to those who had unwittingly fabricated it. The legend of one of Goya's *Caprichos* etchings—"The dream of reason produces monsters"—seemed as much social science as surrealism.

Spengler's famous book popularized one version of the resulting bewilderment, but it rested upon a fragile analogy. In the *Decline of the West,* what was happening had always happened. History, like biology, was moving through the immemorial cycle of birth, maturity, and death. Max Weber was much more profound. What was happening had never happened before. Technological progress was achieved by a radical method of breaking life up into specific functions which could be measured and engineered. In such a subdivided existence, there was no vantage point for the comprehension of the whole.

Bureaucratic, scientific man was losing his intellectual hold on reality even as he pragmatically conquered it.

It is still possible to counterpose a hope to these pessimisms of Spengler and Weber (and their themes are present in most of the contemporary dooms). The accidental revolution could become conscious of itself, and the future would thus be chosen rather than submitted to. To make this hope anything more than a wish, it is necessary to understand the attack that the twentieth-century genius mounted upon it. And this is a main purpose of this optimistic study of decadence.

In attempting such an anatomy of the various destinies assigned to these times, a typical excess of the century emerges at the outset: that it has supported two, seemingly contradictory, apocalypses, one gentle, the other violent.

The profound revolution manifest in Megalopolis came into existence without the intervention of masses in the streets. It can stand as a case of the gentle apocalypse. The violent apocalypse is relatively simple to define. Lenin referred to it when he declared the post-World War I period to be an epoch "of imperialist war and proletarian revolution." Change was to proceed by the road of cataclysm. And between 1914 and 1945, the West went through two world wars (which had the shattering aspect of civil conflict within a common heritage), inflation, a general collapse of the economic system, the rise and fall of fascism, the most bitter and intense class struggle, and, finally, the discovery of a bomb which could annihilate mankind and render the world uninhabitable. It seemed, as one poet put it in September, 1939, that a culture had been driven mad.

In addition to these incredible transformations within the West, the globe itself was in furious change. And though the analysis of this book limits itself to trends within Europe and America, the international context should at least be sketched in as background. Along with all the other reasons for fear and doubt, the West in this period lost its unquestioned rule of the world. In 1914, the white countries were secure in their hegemony over the nonwhite masses of the earth. The inferiority and passivity of the colonial peoples was almost taken for granted. Four or five years later—the turn of the Chinese Revolution toward antiimperialism in May, 1919, is as good a symbolic date as any—that was no longer true. Europe and America, to be sure, retained their political and economic power and hardly saw in a Chinese student demonstration the beginning of a momentous end. But, it became increasingly clearer, some hundreds of years of white dominion were drawing to a close.

This turning point was not, as André Malraux has pointed out, simply political. Up until the reentry of the nonwhite millions into history, it was logical to assume that the cultural standards of the West were as superior as its technology. "Decadence" was anything that departed from the classic, Mediterranean tradition. But then, simultaneously with the loss of Western internal conviction, the formerly subject cultures forced their way rudely into the century. Suddenly, African masks were no longer the fetishes of primitives but an art about demons that had a reference for the advanced nations.

Indeed, the violent apocalypse can be traced in art as well as in history. A writer like Thomas Mann is a typical case in point. The time literally took him by the scruff of his neck, transforming his novels as well as his politics, ultimately leading him to conjure up the Devil in order to explain what his native Germany was doing to itself. Mann was not alone in his shock. The aesthetic revolution had begun even before the First World War, and the Philistines were right for the wrong reasons in considering the radical dissolution of traditional forms as the harbinger of a new relativism.

Between the two wars, the arts were implicated in upheaval like everything else. Dada discovered that tone of inspired mockery that still persists. The Surrealists attempted the wedding of Freud and Marx, dream and technology. There were antonalism, abstraction, the modern dance, and many other departures. If one knew nothing of these years from history, their apocalyptic character could be inferred from their art.

The response of two sensitive men, the one a Czech novelist, the other a British scholar, can stand as a documentation of this point. After World War I, the haunting modern genius, Frank Kafka, remarked, "The war has opened the flood gates of chaos. The buttresses of human existence are collapsing. Historical development is no longer determined by the individual but by the masses. We are shoved, rushed, swept away. We are the victims of history."

And, writing in the 1930's from the very midst of the violent apocalypse, Gilbert Murray said, "There is something wrong. There is a loss of confidence, a loss of faith, an omnipresent, haunting fear. People speak as they never spoke in Victorian days of the possible collapse of civilization." The Victorians, Murray asserted, had a cosmos (it included the nebular hypothesis, the theory of evolution, the liberating potential of science, a sort of Christianity, and a concept of the gentleman); the moderns have a chaos.

In economic and political terms, the climactic moment of the violent apocalypse took place in the 1930's. That decade provides an older, more dramatic, example of the accidental revolution than Megalopolis: the paradox of starvation through glut. The West had achieved the highest technical proficiency in the history of man. There were desperate, crying needs which demanded fulfillment. There were idle resources, material and human, which could have satisfied them. Yet society could not put its industrial capacity and its people to work to meet these obvious necessities. In Germany, the failure led to fascism, and in the rest of the Western world to sharp class conflict. Wealth had labored to produce hunger; empty factories brooded over the men who had made them but could not use them.

Murray's fears are signed with the mood of those times. They are the sentiments of an educated British gentleman but, in different forms, they were shared by people in every Western nation. This was the stuff of the violent apocalypse, the experience that led many to consider the "possible collapse of civilization." At such a time, the idea of decadence was not academic and scholarly, but political. It was an intellectual element in the growth of revolutionary mass movements on the Right and Left, and communists, socialists, and fascists all spoke of the decadence of the bourgeois society.

And yet, it is important to realize that this explosive and unstable period is related to the quiet revolution of Megalopolis. Paul Valéry was one of the first to recognize that there were two decandences possible in the twentieth century. The first would see "a depression of intellectual values, a lowering, a decadence comparable to the one produced at the end of antiquity; culture abandoned, masterpieces become incomprehensible or destroyed . . ." This prophecy was in the mood of Gilbert Murray. But then, Valéry also saw another terrible option: "the application of industrial methods to the production, evaluation and consumption of the fruits of the spirit would end by transforming the highest and most important intellectual virtues . . . " And this is an intimation of the gentle decadence.

At first glance, there is little connection between the thirties and the sixties, between the Depression and class struggle on the one hand, and prosperity and internal stability on the other. Yet, these are two modes of the accidental revolution. In each instance, technology is more thoughtful and creative than men. In the case of the violent apocalypse, the failure to control the work of man's hands led to mass unemployment, fascism, war, a rending of the Western social fabric. This caused many people to talk about revolution, usually without making one. In the gentle apocalypse, Megalopolis and similar developments occurred behind the backs of those whose lives they transformed. People made a revolution without talking about it.

Practically everyone in the West today recognizes that a violent apocalypse took place in the period *entre les deux guerres.* But now, it is said, all that is past, and talk of revolution or basic change is hopelessly outdated. Through modifications in their structure, the advanced nations have brought the economic demons under control and can now attend to the aesthetics of social life. The profundity of Megalopolis challenges such a complacency. But beyond the statistics, it is important to understand the reason behind this refusal to notice that human existence is being remade.

And here again, Friedrich Nietzche is relevant. In an 1888 polemic against Wagner, the German philosopher wrote: "What characterizes the literary decadence? It is that the whole no longer has life. The word becomes sovereign and leaps up out of he sentence, the sentence reaches out and obscures the page, and the page comes to life at the expense of the whole . . ."

Nietzsche's literary judgment can be translated into a description of much of social thought today. In it, "the page comes to life at the expense of the whole." The large ideas about society's alternatives have faded into the background. The tiny increments of change are minutely examined; their sweeping, radical sum is all but ignored. Between the two wars, and particularly in the thirties, this was not the case, mainly because the Nietzschean prediction of upheaval was so obviously true. With the breakdown of the Western economy and the barbaric retrogression in Germany, history was ransacked for explanations. Significant sections of the population debated and struggled over the order of things to come.

In the post-World War II years, this all changed once more. The revolutionary hopes of the European Resistance were cruelly disappointed, the United States did not return to the conditions of the Depression. As a result, some social scientists in the fifties and early sixties talked of the "end of ideology." The old debates had, they said, become sterile. In part, these thinkers were reacting to some of the oversimplifications of the thirties. In part they expressed a general mood of relative economic and political satisfaction.

It is difficult to grasp a revolutionary evolution, which is what the gentle apocalypse is. In the thirties, it was a question of a storm; in the sixties, it is one of a deep sea change below the social surface. And both moments are the expression of the same persistent turbulence of the twentieth century.

Now that a revolution is indeed taking place in every aspect of social, economic, and political life, such concepts as capitalism, socialism, and democracy are relevant. There is an imprecision to such generalities, but there is also an imprecision to sociological minutiae, and at the present a more dangerous one. To attempt to anticipate in outline form the large idea which makes up the main argument of this study: The accidental revolution has resulted, not in this or that loss of faith, but in introducing doubt and contradiction into every Western creed, secular or religious. In time of gentle apocalypse, such as the present, it is possible to ignore such a convergence of crises or to treat them pragmatically. The revolution is going its casual way, there are no Western *coups d'état* to take a position for or against, and one can hope that the situation will be blundered through. Still, it is of some consequence that capitalism, socialism, democracy, religion, and atheism have simultaneously become problematic. This fact might become even more urgent tomorrow if, as can never be discounted in this century, the revolution would once more become revolutionary.

Here are some of the decadences with which the modern world is concerned.

Adam Smith has been stood on his head. In the classic theory, the invisible hand of the market directed the sum total of private and individual decisions toward a common good. This was the element in the Western economic creed that made it something more than a rationalization for greed. According to this view, the market not only resulted in the most efficient use of material resources but strengthened the freedom of choice of the citizen and ultimately created the happiness of the largest possible number of people. Thus, capitalist man was accomplishing moral and political virtue as well as observing economic reason when he vigorously pursued his personal gain.

The mystical equilibrium of this theory was never realized in reality. Yet, in a rough way the activity of the entrepreneurs did raise the level of the entire society. It was, to be sure, necessary for the people, and particularly the workers, to sacrifice and struggle in order to achieve their share of the new wealth (it was not delivered to them by the market mechanism). But there was still some sense to the Smithian equation of the public and private gain.

But the growing divergence, and even conflict, between private and public cost has converted the capitalist economy into the near opposite of Smith's description. Not only do the individual decisions add up to a revolution rather than a harmony, but they often dissatisfy their successful makers. One of the wriest spectacles of the age is that of business men lamenting the "socialistic" drift of the system which has made them richer year by year (or, in the thirties, denouncing the anti-Depression measures which were designed to save them). This amounts to a confused, corporate acknowledgment that the old rules no longer apply.

So capitalism no longer explains the captialist reality. Indeed in more than one sense the most efficient anticapitalists inside the Western world are the businessmen.

Socialism is in a different kind of crisis. In the patristic statements of the nineteenth century, the socialist fathers argued that utopia had become practical. In the past, they said, the good society was a dream of poets and philosophers as well as a persistent, but impossible, instinct of the mass. But with the Industrial Revolution, history turned a corner. For the first time ever, the production of genuine abundance was within the technical competence of man. As a result, there was a material basis for human decency. And with the appearance of the modern working class, driven to economic and political organization by the necessities of daily life, a social force had come into being that would be impelled toward the theory and practice of justice.

For a considerable time, the socialist hope seemed to be coming true. A huge, revolutionary movement grew up in Europe, and Karl Liebknecht could cry out in the German Revolution at the end of World War I, "We are storming the gates of heaven!" But then, the workers in the West did not seize power, confusion and fascism came, and more recently—this is the gentle apocalypse of the labor movement—the changes in the manpower structure of Western society have begun to reduce the absolute and relative number of direct producers. It is now possible to think of a cybernated age that will dispense to a large degree with the blue-collar men on the assembly line, with clerks, and even middle-level executives.

Yet, at the same time that the political will of socialism seemed to falter, the basic argument which had originally inspired the movement grew in force. The means of production, the nineteenth-century socialists had asserted, are social in character, increasingly interdependent, and becoming collective in fact if not in law. But the ownership, management, and guidance of them (and from the point of view of this theory, it makes no difference which of the three preceding terms one chooses to emphasize) remain private. In order to direct this technology humanely, it is necessary to place it under democratic social control. This thesis grows more apt every day. Thus, at that point at which the socialist vision has become all the more compelling, the number of people who take it with political seriousness in the West seems to have declined.

One element in this socialist crisis has implications for the very concept of economics and society itself. The change in the manpower structure of the West, the recent shift (mainly in America) from mass production by men to mass production by machines, threatens some of the culture's most cherished economic, psychological, and ethical assumptions. Immemorially, it has been assumed that economics is concerned with being economical—with the use of scarce resources. It is this premise which has motivated the Western obsession with efficiency. Now, this conventional wisdom, as John Kenneth Galbraith has called it, is less and less descriptive of a society which has the potential to produce more with fewer workers.

Every President of the United States since Harry Truman has proclaimed that it is the duty of the citizen to consume. The statement is usually made in the course of treating the details of economic policy, yet it is fraught with the most important ethical and psychological implications.

Another related crisis is that of democracy. In the nineteenth century, it was thought that the growth in social wealth and literacy would produce a more democratic population with a higher set of values. However, a series of recent analyses have suggested another denouement: that as a result of these changes, the West has created a new kind of mass society in which education and an increased standard of living have the effect of pulling culture down to the lowest common denominator. And there are now those who look back nostalgically to idealized days of aristocracy.

The character of contemporary technology seems to reinforce this anti-democratic trend in modern life. Part of the manpower revolution is the division of the society into the highly skilled and the highly unskilled. With so much economic, political, and social power concentrating in computerized industry, the question arises, who will do the programming? Who will control the machines that establish human destiny in this century? And there is clearly the possibility that a technological elite, perhaps even a benevolent elite, could take on this function.

Profound shifts like these cannot, of course, be confined to the economic and political sphere. Up until the twentieth century, religion had spoken to men who were haunted by plague, famine, and natural disaster. Within the last hundred years, large portions of these traditional domains of God have been mastered by science. It is difficult to pray for rain when the meteorologists can predict it and, in some cases, even precipitate it. So the question arises, where does God live now? With man more and more ubiquitous, with nature transformed from a mysterious given into a product of the human will, divinity is in crisis.

But the confusion of the religionists has not resulted in the triumph of the atheists. In the old calculus of godlessness, the irrationality of religion had been imposed upon man because of his impotence in the face of the world. Once the external environment was understood, man would no longer require the solace of an imagined God. Yet, it has been characteristic of the accidental revolution that the more the various aspects of society have become rationalized, the more the totality of society seems to be inexplicable. If the awesomeness of the world that God was once said to have created has declined, the opacity of the world that man has made has increased. And this new mystery is all the more bewildering since it cannot be blamed on, or justified by, the supernatural.

One could go on listing crises almost indefinitely, yet the main point should be clear by now. The chasm between technological capacity and economic, political, and social, and religious consciousness—the accidental revolution, in short—has unsettled every faith and creed in the West. This has led many people to a sense of decadence. The theories that express this mood relate to every aspect of human life. They embrace psychology, religion, ethics, and art, and there are important things to be seen from each of these specific vantage points. But in a complex way, the accidental revolution is party to every one of these developments.

The result of this process, the summary paradox, is that these most conscious and man—made of times have lurched into the unprecented transformation of human life without thinking about it. And in a sense, this century, this scientific, technological, and utterly competent century, has happened accidentally.

In the description of decadence and the accidental century that follows, I have made no attempt to be neutral. In an age such as this, when change is epidemic, the present does not hold still long enough to be studied with archeological objectivity. Today is always partly tomorrow and can only be understood in movement, futuristically, speculatively. So let me state my bias openly. The hope for the survival and fulfillment of the Western concept of man demands that the accidental revolution be made conscious and democratic. And to argue this requires a restatement—or, perhaps, to borrow a word from Pope John XXIII, an *aggiornamento*—of the socialist ideal and the socialist possiblity.

In *The Accidental Century*, Michael Harrington sets forth his conviction that society has the power and ability to guide the Technological Revolution into courses beneficial to humanity, and that the forces of change can be subjected to rational democratic control. A much more pessimistic and deterministic analysis of modern man's dilemma is given by Jacques Ellul, veteran of the French Resistance and Professor of History and Sociology at the University of Bordeaux. Professor Ellul vehemently rejects the concept that technology is in itself neutral since it is man, choosing the ends to which technology shall be directed, who makes it a force for good or evil. Instead, M. Ellul insists upon the inexorability of "technique"—his term for the various rationalized means which technological society used to achieve the goal of absolute efficiency and total control. The impersonal and relentless forces of "technique," he argues, are ruthlessly imposing standardization, centralization, and efficiency upon society, thus progressively dehumanizing it. Once "technique" has taken hold, it must move irreversibly and uncontrollably, according to its own inner logic, toward its own ends and purposes. Professor Ellul suggests that those ends and purposes may well prove to be quite tragic for modern civilization.

Jacques Ellul on the Characterology of Technique*

*   *   *   *   *

The great tendency of all persons who study techniques is to make distinctions. They distinguish between the different elements of technique, maintaining some and discarding others. They distinguish between technique and the use to which it is put. These distinctions are completely invalid and show only that he who makes them has understood nothing of the technical phenomenon. Its parts are ontologically tied together; in it, use is inseparable from being.

•Source: From THE TECHNOLOGICAL SOCIETY, by Jacques Ellul pp. 94-111 Copyright © 1964 by Alfred A. Knopf, Inc. Reprinted by permission of the publisher. (Sections have been omitted with permission of the Publisher.)

It is common practice, for example, to deny the unity of the technical complex so as to be able to fasten one's hopes on one or another of its branches. Mumford gives a remarkable example of this when he contrasts the grandeur of the printing press with the horridness of the newspaper. "On the one side there is the gigantic printing press, a miracle of fine articulation . . . On the other the content of the papers themselves recording the most vulgar and elementary emotional states . . . There the impersonal, the cooperative, the objective; here the limited, the subjective, the recalcitrant, the ego, violent and full of hate and fear, etc. . . ." Unfortunately, it did not occur to Mumford to ask whether the content of our newspapers is not really necessitated by the social form imposed on man by the machine.

This content is not the product of chance or of some economic form. It is the result of precise psychological and psychoanalytical techniques. These techniques have as their goal the bringing to the individual of that which is indispensable for his satisfaction in the conditions in which the machine has placed him, of inhibiting in him the sense of revolution, of subjugating him by flattering him. In other words, journalistic content is a technical complex expressly intended to adapt the man to the machine.

It is certain that a press of high intellectual tone and great moral elevation either would not be read (and then one would scarcely see the wherefore of these beautiful machines) or would provoke in the long run a violent reaction against every form of technical society, including the machine. This reaction would come about not because of the ideas such a press would disseminate, but because the reader would no longer find in it the indispensable instrument for releasing his repressed passions.

In a sound evaluation of the problem, it ought never to be said: on the one side, technique: on the other, the abuse of it. There are different techniques which correspond to different necessities. But all techniques are inseparably united. Everything hangs together in the technical world, as it does in the mechanical; in both, the advisability of the isolated means must be distinguished from the advisabiltiy of the mechanical "complex." The claims of the mechanical "complex" must prevail when, for example, a machine too costly or overrefined threatens to wreck the ensemble.

There is an attractive notion which would apparently resolve all technical problems: that it is not the technique that is wrong, but the use men make of it. Consequently, if the use is changed, there will no longer be any objection to the technique.

I shall return more than once to this conception. Let us examine a single aspect of it now. First, it mainfestly rests on the confusion between machine and technique. A man can use his automobile to take a trip or to kill his neighbors. But the second use is not a use; it is a crime. The automobile was not created to kill people, so the fact is not important. I know, of course, that killing people is not what those who explain things in this way have in mind. They prefer to say that man orients his pursuits in the direction of good and not of evil. They mean that technique seeks to invent rational therapies and not poison gases, useful sources of energy and not atomic bombs, commercial and not military aircraft, etc. This leads them straight back to man—man who decides in what direction to orient his researches. (Must it not be, then, that man is becoming better?) But all this is an error. It resolutely refuses to recognize technical reality. It supposes, to begin with, that men orient technique in a given direction for moral, and consequently nontechnical, reasons. But a principal characteristic of technique (which we shall study at length) is its refusal to tolerate moral judgments. It is absolutely independent of them and eliminates them from its domain. Technique never observes the distinction between moral and immoral use. It tends, on the contrary, to create a completely independent technical morality.

Here, then, is one of the elements of weakness of this point of view. It does not perceive technique's rigorous autonomy with respect to morals; it does not see that the infusion of some more or less vague sentiment of human welfare cannot alter it. Not even the moral conversion of the technicians could make a difference. At best, they would cease to be good technicians.

This attitude supposes further that technique evolves with some end in view, and that this end is human good. Technique, as I believe I have shown, is totally irrelevant to this notion and pursues no end, professed or unprofessed. It evolves in a purely causal way: the combination of preceding elements furnishes the new technical elements. There is no purpose or plan that is being progressively realized. There is not even a tendency toward human ends. We are dealing with a phenomenon blind to the future, in a domain of integral causality. Hence, to pose arbitrarily some goal or other, to propose a direction for technique, is to deny technique and divest it of its character and its strength.

There is a final argument against this position. It was said that the use made of technique is bad. But this assertion has no meaning at all. As I have pointed out, a number of uses can always be made of the machine, but only one of them is the technical use. The use of the automobile as a murder weapon does not represent the technical use, that is, the one best way of doing something. Technique is a means with a set of rules for the game. It is a "method of being used" which is unique and not open to arbitrary choice; we gain no advantage from the machine or from organization if it is not used as it ought to be. There is but one method for its use, one possibility. Lacking this, it is not a technique. Technique is in itself a method of action, which is exactly what a use means. To say of such a technical means that a bad use has been made of it is to say that no technical use has been made of it, that it has not been made to yield what it could have yielded and ought to have yielded. The driver who uses his automobile carelessly makes a bad use of it. Such use, incidentally, has nothing to do with the use which moralists wish to ascribe to technique. Technique *is* a use. Moralists wish to apply another use, with other criteria. What they wish, to be precise, is that technique no longer be technique. Under the circumstances, there are no further significant problems.

There is no difference at all between technique and its use. The individual is faced with an exclusive choice, either to use the technique as it should be used according to the technical rules, or not to use it at all. It is impossible to use it otherwise than according to the technical rules.

Unfortunately, men today accept this reality only with difficulty. Thus, when Mumford makes the statement: "The army is the ideal form towards which a purely mechanical industrial system must tend," he is unable to restrain himself from adding: "But the result is not ideal." What is the "ideal" doing here? The ideal is not the problem. The problem is solely to know whether this mode of organization responds to technical criteria. Mumford is able to show that it is nothing of the kind, because he limits techniques to machines. But if he were to accept the role of human techniques in the organization of the army he could account for the fact that the army indeed remains the irreproachable model of a technical organization, and its value has nothing to do with an ideal. It is infantile to wish to submit the machine to the criterion of the ideal.

It is also held that technique could be directed toward that which is positive, constructive, and enriching, omitting that which is negative, destructive, and impoverishing. In demagogic formulation, techniques of peace must be developed and techniques of war rejected. In a less simple-minded version, it is held that means ought to be sought which palliate, without increasing, the drawbacks of technique. Could not atomic engines and atomic power have been discovered without creating the bomb? To reason thus is to separate technical elements with no justification. Techniques of peace and alongside them other and different techniques of war simply do not exist, despite what good folk think to the contrary.

The organization of an army comes to resemble more and more that of a great industrial plant. It is the technical phenomenon presenting a formidable unity in all its parts, which are inseparable. The fact that the atomic bomb was created before the atomic engine was not essentially the result of the perversity of technical men. Nor was it solely the attitude of the state which determined this order. The action of the state was certainly the deciding factor in atomic research (I shall take up this point later). Research was greatly accelerated by the necessities of war and consequently directed toward a bomb. If the state had not been oriented toward the ends of war, it would not have devoted so much money to atomic research. All this caused an undeniable factor of orientation to intervene. But if the state had not promoted such efforts, it would have been the whole complex of atomic research which would have been halted without distinction between the uses of war and peace.

If atomic research is encouraged, it is obligatory to pass through the stage of the atomic bomb; the bomb represents by far the simplest utilization of atomic energy. The problems involved in the military use of atomic energy are infinitely more simple to resolve than are those involved in its industrial use. For industrial use, all the problems involved in the bomb must be solved, and in addition certain others, a fact corroborated by J. Robert Oppenheimer in his Paris lecture of 1958. The experience of Great Britain between 1955 and 1960 in producing electricity of nuclear origin is very significant in this respect.

It was, then, necessary to pass through the period of research which culminated in the bomb before proceeding to its normal sequel, atomic motive power. The atomic-bomb period is a transitory, but unfortunately necessary, stage in the general evolution of this technique. In the interim period represented by the bomb, the possessor, finding himself with so powerful an instrument, is led to use it. Why? Because everything which is technique is necessarily used as soon as it is available, without distinction of good or evil. This is the principal law of our age. We may quote here Jacques Soustelle's well-known remark of May, 1960, in reference to the atomic bomb. It expresses the deep feeling of us all: "Since it was possible, it was necessary." Really a master phrase for all technical evolution.

Even an author as well disposed toward the machine as Mumford recognizes that there is a tendency to utilize all inventions whether there is need for them or not. "Our grandparents used sheet iron for walls although they knew that iron is a good conductor of heat . . . The introduction of anesthetics led to the performance of superfluous operations . . ." To say that it could be otherwise is simply to make an abstraction of man.

Another example is the police. The police have perfected to an unheard of degree technical methods both of research and of action. Everyone is delighted with this development because it would seem to guarantee an increasingly efficient protection against criminals. Let us put aside for the moment the problem of police corruption and concentrate on the technical apparatus, which, as I have noted, is becoming extremely precise. Will this apparatus be applied only to criminals? We know that this is not the case; and we are tempted to react by saying that it is the *state* which applies this technical apparatus without discrimination. But there is an error of perspective here. The instrument tends to be applied *everywhere* it *can* be applied. It functions without discrimination—because it exists without discrimination. The techniques of the police, which are developing at an extremely rapid tempo, have as their necessary end the transformation of the entire nation into a concentration camp. This is no perverse decision on the part of some party or government. To be sure of apprehending criminals, it is necessary that *everyone* be supervised. It is necessary to know exactly what every citizen is up to, to know his relations, his amusements, etc. And the state is increasingly in a position to know these things.

This does not imply a reign of terror or of arbitrary arrests. The best technique is one which makes itself felt the least and which represents the least burden. But every citizen must be thoroughly known to the police and must live under conditions of discreet surveillance. All this results from the perfection of technical methods.

The police cannot attain technical perfection unless they have total control. And, as Ernst Kohn-Bramstedt has remarked, this total control has both an objective and a subjective side. Subjectively, control satisfies the desire for power and certain sadistic tendencies. But the subjective aspect is not the dominant one. It is not the major aspect, the expression of what is to come. In reality, the objective aspect of control—more and more, that is to say, the pure technique which creates a milieu, an atmosphere, an environment, and even a model of behavior in social relations—dominates more and more. The police must move in the direction of anticipating and forestalling crime. Eventually intervention will be useless. This state of affairs can come about in two ways: first, by constant surveillance, to the end that noxious intentions be known in advance and the police be able to act before the premeditated crime takes place; second, by the climate of social conformity which we have mentioned. This goal presupposes the paternal surveillance of every citizen and, in addition, the closest possible tie-in with all other techniques—administrative, organizational, and psychological. The technique of police control has value only if the police are in close contact with the trade unions and the schools. In particular, it is allied with propaganda. Wherever the phenomenon is observed, this connection exists. Propaganda itself cannot be efficient unless it brings into play the whole state organization, and particularly the police power. Conversely, police power is a genuine technique only when it is supplemented by propaganda, which plays a leading role in the psychological environment necessary to the completeness of the police power. But propaganda must also teach acceptance of what the police power is and what it can do. It must make the police power palatable, justify its actions, and give it its psychosociological structure among the masses of the people.

All this is equally true for dictatorial regimes in which police and propaganda concentrate on terror, and for democratic regimes in which the motion pictures, for example, show the good offices of the police and procure it the friendly feeling of the public. The vicious circle mentioned by Ernst Kohn-Bramstedt (past terror accentuates present propaganda, and present propaganda paves the way for future terror) is as true of democratic as of dictatorial regimes, if the term *terror* is replaced by *efficiency*.

This type of police organization is not an arbitrary prospect. It is maintained by every authoritarian government, where every citizen is regarded as a suspect ignorant of his own capabilities. It is the tendency in the United States, and we are beginning to see the first elements of it in France. The administration of the

French police was oriented, in 1951, toward an organization of the system "in depth." This took place, for example, at the level of the Record Office. Certain elements of this are simple and well known: fingerprint files, records of firearms, application of statistical methods which allow the police to obtain in a minimum of time the most varied kinds of information and to know from day to day the current state of criminality in all its forms. Other elements are somewhat more complicated and new. For example, a punched-card mechanical index system (*Recherches*) has been installed in the Criminal Division. This system offers four hundred possible combinations and permits investigations to begin with any element of the crime: hour of commission, nature, objects stolen, weapons used, etc. The combination obviously does not give the solution but a series of approximations.

The most important item in this catalogue of police techniques is the creation of the so-called "suspect files", which show whether the police ever suspected any individual for any reason or at any time whatsoever, even though no legal document or procedure ever existed against him (from the press conference of M. Baylot, Prefect of Police, 1951). This means that any citizen who, once in his life, had anything to do with the police, even for noncriminal reasons, is put under observation—a fact which ought to affect, speaking conservatively, half the adult male population. It is obvious that these lists are only a point of departure, because it will be tempting, as well as necessary, to complete the files with all observations which may have been collected.

Finally, this technical conception of the police supposes the institution of concentration camps, not in their dramatic aspects, but in their administrative aspects. The Nazi's use of concentration camps has warped our perspectives. The concentration camp is based on two ideas which derive directly from the technical conception of the police: preventive detention (which completes prevention), and re-education. It is not because the use of these terms has not corresponded to reality that we feel it necessary to refuse to see in the concentration camp a very advanced form of the system. Nor is it because the so-called methods of re-education have, on the whole, been methods of destruction that we feel we must consider such a concept of "re-education" an odious joke. The further we advance, the more will the police be considered responsible for the re-education of social misfits, a goal that is a part of the very order which they are charged with protecting.

We are experiencing at present the justification of this development. It is not true that the perfection of police power is the result of the state's Machiavellianism or of some transitory influence. The whole structure of society implies it, of necessity. The more we mobilize the forces of nature, the more must we mobilize men and the more do we require order, which today represents the highest value. To deny this is to deny the whole course of modern times. This order has nothing spontaneous in it. It is rather a patient accretion of a thousand technical details. And each of us derives a feeling of security from every one of the improvements which make this order more efficient and the future safer. Order receives our complete approval; even when we are hostile to the police, we are, by a strange contradiction, partisans of order. In the blossoming of modern discoveries and of our own power, a vertigo has taken hold of us which makes us feel this need to an extreme degree. After all, it is the police who are charged, from the external point of view, with insuring this order which covers organization and morals. How then can we possibly deny to the police indispensable improvements in their methods?

We in France are still in the preparatory phase of this development, but the organization of police power has been pushed very far in Canada and New Zealand, to take two examples. Technical necessity imposes the national concentration camp (which, I must point out, does not involve the suffering usually associated with it).

Let us take another example. A new machine of great productive power put into circulation "releases" a great quantity of work; it replaces many workers. This is an inevitable consequence of technique. In the crude order of things, these workers are simply thrown out of work. Capitalism is blamed for this state of affairs and we are told that technique itself is not responsible for technological unemployment and that the establishment of socialism would set things right. The capitalist replies: "Technological unemployment always dies out of itself. For example, it creates certain new activities which will in the long run create employment for qualified workers." This appears to be a dreadful prospect because it implies a readaptation in *time* and more or less lengthy period of unemployment. But what does socialism propose? That the "liberated" worker will be used somewhere else and in some other capacity. In the Soviet Union the worker is either adapted to a new skill by means of vocational training or he is sent to another part of the country. In the Beveridge Plan the worker is employed wherever the state opens a plant of any sort. This socialist solution involves readaptation in space. But this solution, too, appears to be completely alien to human nature. Man is not a mere package to be moved about, an object to be molded and applied wherever there is need. These two forms of readaptation, the only ones possible, are both inhumane. The New Work Code promulgated in the (East) German Democratic Republic in November, 1960, shows this inhumanity in operation in the socialist camp. And none of these adaptations can be separated from the machine which replaces human labor. They are its necessary and inevitable consequence. Of course, idealists will speak of the reduction of the work week. But this reduction can only be effected when equivalent technical improvements are produced in all fields of work. According to Colin Clark, it seems that this reduction, too, must "ceiling out" before long. But this consideration passes over into the area of economics.

I could cite innumerable examples, but the ones I have given suffice to show that technique in itself (and not the use made of it, or its non-necessary consequences) leads to a certain amount of suffering and to social scourges which cannot be completely separated from it. This is its very mechanism.

Of course, a technique can be abandoned when it proves to have evil effects which were not provided for. From then on, there will be an improvement in the technique. A characteristic example is furnished by J. de Castro in *The Geography of Hunger*. De Castro shows in detail, with regard to Brazil, what was already known superficially about other countries, that certain techniques of exploitation have proved disastrous. According to de Castro, certain regions were deforested in order to grow sugar cane. But only the immediate technical productivity was considered. In a further work, de Castro seeks to show that the hunger problem was created by application of the capitalist and colonialist system to agriculture. His reasoning, however, is correct only to a very limited extent. It is true that when an agriculture of diversified crops is replaced by a single-crop economy for commercial ends (tobacco and sugar cane), capitalism is to blame. But most often crop diversification is not disturbed. What happens is that new areas are brought under cultivation, producing a population increase and also a unilateral utilization of the labor forces. And this is less a capitalist than a technical fact. If the possibility of industrializing agriculture exists, why not use it? Any engineer, agronomist, or economist of a hundred years ago would have agreed that bringing uncultivated lands under cultivation constituted a great advance. The application of European agricultural techniques represented an incomparable forward step, when compared, for example, to Indian methods. But it involved certain unforeseen consequences: the resulting deforestation modified hydrographic features, the rivers became torrents, and the drainage waters provoked catastrophic erosion. The topsoil was completely carried away and agriculture became impossible. The fauna, dependent of the existence of the forest, disappeared. In this way, the food-producing possibilities of vast regions vanished. The same situation is developing as a result of the cultivation of peanuts in Senegal, of cotton in the South of the United States, and so on. None of this represents, as is commonly said, a poor application of technique—one guided by selfish interest. It is simply technique. And if the situation is rectified "too late" by the abandonment of the old technique, it will only be as a consequence of some new technical advance. In any case, the first step was inevitable; man can never foresee the totality of consequences of a given technical action. History shows that every technical application from its beginnings presents certain unforeseeable secondary effects which are much more disastrous than the lack of the technique would have been. These effects exist alongside those effects which were foreseen and expected and which represent something valuable and positive.

Technique demands the most rapid possible application; the problems of our day are evolving rapidly and require immediate solutions. Modern man is held by the throat by certain demands which will not be resolved simply by the passage of time. The quickest possible counter-thrust, often a matter of life or death, is necessary. When the parry specific to the attack is found, it is used. It would be foolish not to use the available means. But there is never time to estimate all the repercussions. And, in any case, they are most often unforeseeable. The more we understand the interrelation of all disciplines and the interaction of the instruments, the less time there is to measure these effects accurately.

Moreover, technique demands the most immediate application because it is so expensive. It must "pay off," in money, prestige, or force (depending on whether the regime is capitalist, Communist, or Fascist, respectively). There is no time for precautions when the distribution of dividends or the salvation of the proletariat is at stake. Nor can we permit ourselves to say that these motives are no affair of technique. If none of them existed, there would be no money for technical research and there would be no technique. Technique cannot be considered in itself, apart from its actual modes of existence.

We are brought back, then, to serious facts of this order: in certain agricultural research in England, antiparasitic agents called *systemics* were applied. An injection was made into a fruit tree, which as a consequence was infected with the agent from its roots to its leaves. Every parasite died. But nothing is known of the effects on the fruit, or of the effects on man, and *in the long run* of the effects on the tree. All that is known is that the agent is not an immediate deadly poison for the consumer. Such products are already commercially available, and it is probable that they will shortly be used on a large scale. What we have said about systemics holds for the specific insecticide, D. D. T. It was announced originally that this insecticide was completely harmless for warm-blooded animals. Subsequently, D. D. T. was widely used. But it was noted in 1951 that D. D. T. in fatty solution (oily or otherwise) is actually a poison for warm-blooded animals and causes a whole complex of disturbances and diseases, in particular, rickets. This fatty solution may be produced entirely by accident, as when cows treated with this chemical produce milk containing D. D. T. Rickets have been detected in calves nourished with such milk. And several international medical congresses since 1956 have drawn attention to the grave danger to children.

But the real question is not the question of error. Errors are always possible. Two facts alone concern us: it is impossible to foresee all the consequences of a technical action; and technique demands that everything it produces be brought into a domain that affects the entire public.

The weight of technique is such that no obstacle can stop it. And every technical advance is matched by a negative reverse side. An excellent study of the effect of petroleum explorations in the Sahara (1958) concludes with the observation that the most serious problem is the increase in the wretchedness of the local population. The causes of this growing misery, among others, are: the supplanting of caravan traffic by motor vehicles; the disappearance of the date palms (diseased through widespread chemical wastes); and the disappearance of cereal grains because of nonmaintenance of the irrigation works. This complex seems to represent a typical example.

The human being is delivered helpless, in respect to life's most important and most trivial affairs, to a power which is in no sense under his control. For there can be no question today of man's controlling the milk he drinks or the bread he eats, any more than of his controlling his government. The same holds for the development of great industrial plants, transport systems, motion pictures, and so on. It is only after a period of dubious experimentation that a technique is refined and its secondary consequences are modified through a series of technical improvements. Henceforth, someone will say, it will be possible to tame the monster and separate the good results of a technical operation from the bad. That may be. But, in the same framework, the new technical advance will in its turn produce further secondary and unpredictable effects which are no less disastrous than the preceding ones (although they will be of another kind). De Castro declares that the new techniques of soil cultivation presuppose more and more powerful state control, with its police power, its ideology, and its propaganda machinery. This is the price we must pay.

William Vogt, surveying the same problem, is still more precise: in order to avoid famine, resulting from the systematic destruction of the topsoil, we must apply the latest technical methods. But conservation will not be put into practice spontaneously by individuals; yet, these methods must be applied globally or they will not amount to anything. Who can do this? Vogt, like all good Americans, asserts that he detests the authoritarian police state. However, he agrees that only state controls can possibly produce the desired results. He extols the efforts made by the liberal administration of the United States in this respect, but he agrees that the United States continues "to lose ground literally and figuratively," simply because the methods of American agricultural administration are not authoritarian enough.

What measures are to be recommended? The various soils must be classified as to possible ways to cultivate them without destroying them. Authoritarian methods must be applied in order (a) to evacuate the population and to prevent it from working the imperiled soil; and (b) to grow only certain products on certain types of soil. The peasant can no longer be allowed freedom in these respects. This evolution is to be facilitated by centralization of the great land holdings. In Latin America there are today from 20 to 40 million ecologically displaced persons, persons occupying lands which ought not to be under cultivation. They are living on hillsides from which it is absolutely necessary to drive them if the means of existence of their countries are to be saved from destruction. It will be difficult and costly to relocate these people, but Latin

America has no choice. If she does not solve this problem, she will be reduced to the most miserable standard of living.

All experts on agricultural questions are in fact in fundamental agreement. De Castro (although hostile to the ideas of Vogt) and Dumont (critical of de Castro on certain points) come to the conclusion that only strict planning on a world scale can solve the problems of agriculture, and that only human relocation and collective distribution of wealth can solve the problem of famine. This can only mean that man, if he is to improve the traditional agricultural techniques and be rid of their drawbacks, will be obliged to apply extremely rigorous administrative and police techniques. Here again we have a good example to the interconnection of different elements and of the unpredictability of the secondary effects.

It was believed for a long time that the TVA was a praiseworthy response to certain problems raised by technique. Today, however, certain major flaws have become apparent. For example, the correct application of methods of reforestation and animal reproduction were not understood. Flood control was not carried out by retention of the water in the soil but by submerging permanently a good part of the lands which have been saved to protect others. Man, we repeat, is never able to foresee the totality of effects of his technique. No one could have foreseen that regulating the Colorado River for irrigation purposes would lead the Pacific Ocean to encroach upon the coast of California, or that it would endanger the valleys (which had been "regulated") by the removal of up to 500 tons a day of sand and rock. It is likewise impossible to foresee the effect of techniques intended to control the weather, dispel clouds, precipitate rain or snow, and so on. In another area, Professor Lemaire, in a study of narcotic drugs, shows that technique permits the manfacture of synthetic narcotics with greater and greater ease and in increasing quantities. But, according to Lemaire, the control of these drugs is thereby rendered more and more difficult because "we cannot predict whether they will or will not be dangerous. The only proof is their habitual use by addicts. But to obtain this proof requires years of experience."

There is scarcely need to recall that universal famine, the most serious danger known to humanity,[3] is caused by the advance of certain medical techniques which have brought with them good and evil inextricably mixed. This is not a question of good or bad use. No more so is the problem, posed by atomic techniques, of the disposal of atomic waste. Atomic explosions are not the real problem. The real problem continues to be that of the disposal of the ceaselessly accumulating waste materials, despite the reassuring but unfortunately partisan explanations of some atomic scientists. The International Agency for Atomic

---

[3] That this problem can be solved seems doubtful to most recent congresses, the Vevey Congress of 1960 among them.

Energy recognized, in 1959, that these wastes represent a deadly peril and that there is no sure way of avoiding it, except perhaps by means of the difficult process of "vitrification" being undertaken in Canada. And all this involves the *peaceful* use of the atom!

In every case, what can really be foreseen more or less clearly is the need of state intervention to control the effects of technical applications. But by the time a technique is modified in the light of these effects, the evil has already been done. When it is proposed to "choose" between effects, it is always too late. It is doubtless still possible to modify any given element, but only at the price of secondary repercussions. Again, it is doubtless possible to produce, by means of rational exploitation of natural resources, enough food to nourish five billion human beings. But this can be accomplished only at the price of forced labor and a new kind of slavery. Whatever point we choose to examine, we always perceive this interrelation of techniques. In 1960, the World Congress for the Study of Nutrition considered the problem of how modern nutrition is vitiated by the use of chemical products which are themselves significant contributory causes of the so-called diseases of civilization (cancer, cardio-vascular illnesses, etc.). But the Congress's studies indicate that the solution can no longer be a return to a "natural" nutrition. On the contrary, a futher step must be taken which involves *completely* artificial alimentation, so-called rational alimentation. It will not be sufficient merely to control grains, meat, butter, and so forth. The stage at which this would have been feasible has been passed. New technical methods must be found. But can we be assured that this new alimentation will in its turn present no danger?

Every rejection of a technique judged to be bad entails the application of a new technique, the value of which is estimated from the point of view of efficiency alone. But we are always unaware of the more remote repercussions. History shows us that these are seldom positive, at least when we consider history as a whole instead of contenting ourselves with examining disconnected phenomena such as the population increase, the prolongation of the average life span, or the shortening of the work week. These are symptoms which perhaps would have meaning if man were merely an animal, but which have no conclusive significance if man is something more than a production machine.

However, it is not my intention to show that technique will end in disaster. On the contrary, technique has only one principle: efficient ordering. Every-thing, for technique, is centered on the concept of order. This explains the development of moral and political doctrines at the beginning of the nineteenth century. Everything which represented an ordering principle was taken in deadly earnest. At the same time the means destined to elaborate this order were exploited as never before. Order and peace were required for the development of the individual techniques (after society had reached the necessary stage of disintegration). Peace is indispensable to the triumph of industrialization. It will be hastily concluded from this that industrialization will promote peace. But, as always, logical deductions falsify reality. J. U. Nef has shown admirably that industrialization cannot act otherwise than to promote wars. This is no accident, but rather an organic relation. It holds not only because of the direct influence of industrialization on the means of destruction but also because of its influence on the means of existence. Technical progress favors war, according to Nef, because (a) the new weapons have rendered more difficult the distinction between offense and defense; and (b) they have enormously reduced the pain and anguish implied in the act of killing.

On another plane, the distinction between peaceful industry and military industry is no longer possible. Every industry, every technique, however humane its intentions, has military value. "The humanitarian scientist finds himself confronted by a new dilemma: Must he look for ways to make people live longer so that they are better able to destroy one another?" Nef has described all this remarkably well. It is no longer a question of simple human behavior, but of technical necessity.

The technical phenomenon cannot be broken down in such a way as to retain the good and reject the bad. It has a "mass" which renders it monistic. To show this we have taken only the simplest, and hence the most easily debatable, examples. To enable the reader to grasp fully the reality of this monism, it would be necessary to present every problem with all its implications and ramifications into other fields. The case of the police, for example, cannot be considered merely within its specific confines; police technique is closely connected with the techniques of propaganda, administration, and even economics. Economics demands, in effect, an increasing productivity; it is impossible to accept the nonproducers into the body social—the loafers, the coupon-clippers, the social misfits, and the saboteurs—none of these have any place. The police must develop methods to put these useless consumers to work. The problem is the same in a capitalist state (where the Communist is the saboteur) and in a Communist state (where the saboteur is the internationalist in the pay of capitalism).

The necessities and the modes of action of all these techniques combine to form a whole, each part supporting and reinforcing the others. They constitute a co-ordinated phenomenon, no element of which can be detached from the others. It is an illusion, a perfectly understandable one, to hope to be able to suppress the "bad" side of technique and preserve the "good." This belief means that the essence of the technical phenomenon has not been grasped.

"Today, General Electric sailed a ship across the Atlantic, rolled steel in Mexico and turned on the lights in Spain."

General Electric advertisement, 1971

VII. As the Future Unfolds . . .

The final article concerns some of the global implications of those social and technological changes ushered in by the "Technetronic Era." During the 1960's, America's spectacular accomplishments in research and development caused great consternation among certain European observers. Such diverse figures as Prime Minister Wilson of England, France's General DeGaulle, and the influential French journalist, Jean-Jacques Servan-Schreiber warned that Europe was about to become an economic and scientific satellite of the United States. Servan-Schreiber's popularized work, *The American Challenge*, became a best-seller on both sides of the Atlantic. In an essay that could well have been entitled "The American Challenge Revisited, 1967-1970," Dr. Walter Goldstein examines recent developments in the economic relationship of Europe and the United States. Professor Goldstein, an expert in international affairs, particularly European matters, believes that the trends and tendencies he describes are amenable to manipulation and direction by intelligent statesmen. One must also take into account, however, Professor Jacques Ellul's contention that in such matters of "Technique" we are dealing with predetermined forces which cannot be guided or controlled. Finally, Dr. Goldstein raises an issue of long-term importance for diplomacy: how will our runaway technology influence the nature of international relations and affect America's position in world affairs?

231

Walter Goldstein on the Technology Gap*

A new complaint against the United States' role in world affairs has been heard in Europe in the last few years. Though it has not created as great a furor as the war in Vietnam, it has provoked a profound anxiety among the industrial nations of the Western world. Industrialists and economists complain about the grave lag in Europe's rate of economic growth and industrial modernization, though they conceal their fears about the future with facile references to the "technology gap." American business and political leaders have discovered for themselves that the gap between the growth rates of Europe and the United States has widened so dramatically that it is unlikely to be closed in this century. The revolutionary transformation of the productive capacities and technological skills of the American economy has been inadequately recognized within the United States and poorly emulated in Europe. The impact of an unprecedented, advanced, radical technology in American industry, education, and information-handling has left Europe so far behind that a sense of emergency has seized hold in London and in the capital cities of the Common Market. Europe is threatened with a permanent state of inferiority.

Few critics are sure that they know what the term technology gap really means. The gap itself is readily acknowledged; but agreement over its causes, its characteristics, or its consequences has not emerged. A few ideologues have used the term to prove that there is an American master plan for the industrial colonizing of the Continent; others have come to regard it fatalistically as a symbol of Europe's inevitable decline. General de Gaulle played upon this crisis of identity in Europe by insisting that American technology in a dominant role would relegate European industry to a satellite status. To improve Britain's case for entering the Common Market, Harold Wilson warned that America's economic ascendancy would ultimately reduce Europe's industrial plant to a "helotry" holding of the off-shore subsidiaries of the giant United States conglomerates. And the unpredictable Communist leadership in Italy, troubled by the nation's unpromising future, expressed its concern that the influx of American capital and know-how might eventually undermine the viability of Western Europe's capitalist economy.

American discussion about the technology gap has been equally confused but less embittered. Some Americans dismiss it simply as a political slogan aimed at the undermining of our investment and trade in Europe; many insist that it

*Source: Walter Goldstein, "Europe Faces the Technology Gap," *The Yale Review,* Winter 1970 pp. 161-178. Copyright © Yale University. Reprinted by permission.

reflects the protectionist thought and the economic nationalism of a Continent that has failed to modernize its industrial capabilities. Others, more sympathetically inclined, regard it as an expression of a cultural lag, a simple failure of entrepreneurial drive and scientific progress. They fail to realize that the basic future of Europe's economic and social life has been exposed to attack in the last decade by the striking success achieved by American enterprise in penetrating Europe's most promising growth industries.

Though it is difficult to propose an exact definition of the gap between American and European levels of technological innovation, it is essential to explore the European problems that have created the gap. These problems include the massive impact of American capital and management on Europe's most highly prized, science-based industries; a worrisome reliance upon American research and development (R&D) technology in the European thrust to accelerate economic growth; a brain drain of significant dimensions as European professional and scientific workers emigrate; a radical transformation in the archaic management and labor relations of European industry; and a fear—articulated in Jean-Jacques Servan-Schreiber's analysis, *The American Challenge*—that Europeans "are witnessing the prelude to our own historical bankruptcy."

The influx of American investment into Europe's capital structure has been so overwhelming that the fourth largest economy in the world today is that owned by United States corporations operating overseas. In addition, our Gross National Product (GNP) has grown so rapidly in recent years that it now exceeds the combined wealth of all other nations in the competitive market for manufactured goods. A comparison of current GNP figures demonstrates that the economic power of the United States is so vast that potentially it can control the pace of world-wide expansion and modernization. A recent estimate of GNP assessed the worth of our economy (in billions of dollars at current prices) at more than $850; by contrast, the worth of the Soviet Union was $400, West Germany $120, Japan $115, Britain $109, and France $109. Moreover, in per capita terms, an American citizen enjoyed twice as much wealth as the Western European, while the value of United States industrial sales, royalties, and assets in Europe alone surpassed the GNP of the United Kingdom or France.

Though there is a smaller population in the United States than in Western Europe it has become apparent in these post-Cold War years that the difference in economic growth potential is enormous. Last year, with an annual growth rate of almost 5 percent, the United States added nearly as much to its GNP as Canada could claim in total economic wealth. We spent twice as much on improving manufacturing techniques and industrial research as the fourteen North Atlantic Treaty Organization allies devoted to their collectvie defense efforts. Though this energetic growth rate has recently begun to decline, there is little chance that the six members of the Common Market and the seven nations in the European Free Trade Association—or that any larger grouping of Continental economies—will catch up within the century with the industrial impetus and the technological capabilities acquired in the last ten years by the United States.

Another cause of European fears stems from the pattern of distribution of American investments overseas. American capital has concentrated so heavily in the key areas of each economy that Europe's autonomy seems to be dwindling. In 1950, United States capital overseas amounted to less than $20 billion; by 1960 it had reached $50 billion and today it exceeds $100 billion. This growth rate has been three times faster than the average increase in GNP recorded in Western Europe. Though a major share of this investment had once flowed to Canada or Latin America an increasing proportion now flows into Western Europe, antagonizing former Latin friends as well as the new recipients. Cursed with a Midas touch of turning its foreign relations into profit, the American holdings in Europe even managed to survive the severe capital restrictions imposed as a result of the serious outflow of gold incurred because of the Vietnam war.

The unprecedented expansion of our economic empire has deflected one-quarter of our total export trade to subsidiaries of American corporations overseas. The consequences have been dramatic. The United States has utilized its capital and information exports to buy a major if not dominant position in many of the strategic growth or "high technology" industries of Western Europe. These industries provide the critical skills and the technical innovation which every modernizing economy must acquire; and they also provide the best rate of return upon risk-bearing investment. It is not surprising, therefore, that European concern has been intensified there as one giant American corporation after another—or their numerous European subsidiary companies—seized a major role in automobile markets, in the oil and chemical industries, in aircraft and aerospace, and in the manufacture of computers and electronic processing equipment. To the fatalists, at least, the pattern for Europe's future development is clearly evident. As the Old World belatedly adjusts to a second industrial revolution, based upon automated production technology and sophisticated R & D, it is forced to rely more and more heavily upon American know-how, American investment, American industrial licenses, and American management decisions.

The impact exerted on the developed world by American innovations is often revolutionary. While the "Coco-Colonization" of Europe during the 1950's was often the subject of ridicule, the contemporary emphasis on emulation of our industrial mangement and R & D strategy has disrupted many of the tradition-bound or family firms of the Continent. Nowhere has this emulation been more pronounced than in control of the new industries which have burgeoned around each European capital. But the organizations of industry and commerce in the Old World enjoy little of the scientific largesse or the lavish military R & D which American competitors obtain from the United States Government. The aerospace industry in Europe has never been able to match the formidable military and civil expansion programs of Boeing or Lockheed and it may yet lose the race to equip Europe's nationally owned airlines with the newest jet equipment—for example, the intercity airbus or supersonic transport (SST). Though Britain and France have made a desperate attempt to merge their biggest aircraft firms in order to stay ahead in the SST race, their chances for success are not impressive, simply because Boeing has been able to call upon a massive subsidy from the United States Treasury to develop a faster and bigger model. In addition, the outcome of the scramble among British, French, and German firms to develop an airbus design will be determined by the degree of cooperation which the United States airframe industry will provide to each competing nation. Lacking the military financing and the contract pyramids of our aerospace research—which has brought untold wealth to American conglomerates in the airframe and communication satellite industries—Europe now faces a gloomy future of shrinking markets and inadequate technology. Though many striking and original inventions—the jet engine, the vertical takeoff, and the variable geometry aircraft—were pioneered in Britain, the industrial development of these designs quickly passed to larger and better organized American firms. British and French firms now demand their governments' approval for mergers and subsidies, but their production plans are not sufficiently capitalized nor technically advanced enough to exploit the lucrative markets forecast for the 1970's.

Americans have also led in the expanding market for electronic data-processing equipment. International Business Machines and other firms have succeeded in cornering three-quarters of Europe's computer sales, and they are heavily involved in the operations of their remaining Continental competitors. General de Gaulle tried to arrest this trend when he realized that the *force de frappe* and the French space program would rely critically upon American computers. But Machines Bull, the largest French firm, was forced to merge with General Electric in order to remain liquid, and the General's efforts had to be abandoned. A striking fact about this competition is that the equipment patented by European firms has not been inferior to that manufactured by American competitors. The European firms failed in their unimaginative exploitation of new markets and in their inadequate methods of developing their original designs. Today 90 percent of the microcircuitry and computer software used in Europe is of American design or manufacture.

New methods of management and production accompany American innovations in quality control, labor-management negotiation, and forward budget-planning. New concepts of industrial location, training programs, performance budgeting, and market leadership by-pass the archaic procedures of the family firm or the conservative corporation. As John Kenneth Galbraith noted in *The New Industrial State*, the corporate "technostructure" of United States industry has established a powerful autonomy. Unlike the legally hamstrung corporations in the European economy, it has a dominant position in price leadership and in capital formation and has thus an almost oligarchic command over the most lucrative growth sectors of our economy. In repeating this success story in Europe, unfortunately, our industries and their multinational subsidiaries have been so effective in cornering growth markets that they have deflected capital and modernizing skills away from the less flexible, less glamorous industries, such as coal and steel, which must be fully supported if each economy is to maintain a balanced and durable growth. This concentration of investment has brought an excellent rate of return for American capital but it has provoked a severe imbalance in economic development and a marked shortage of domestic capital in the industrial structures of Western Europe.

At present the United States provides one-half of all the foreign capital invested in France, one-third of that in West Germany, and almost three-quarters of all the foreign holdings in Britain. (Indeed, our holdings in Britain increased by nearly a billion dollars in 1967, and in 1968 the United Kingdom was conspicuously excluded from the list of countries from which United States capital was temporarily restricted in order to correct the imbalance of payments caused by the Vietnam war.) Armed with a continuous increase in operating capital, American firms or subsidiaries have succeeded in cornering 30 percent of the market in Europe for automobiles, nearly 30 percent in petrochemical products, and a sizeable percentage in electrical engineering, machine tools, electronic equipment, and aerospace manufactures. Though American investment amounts to only 4 percent of the capital in use in Europe, it has been heavily concentrated in those industries where markets and profits can be most readily expanded. Economic surveys undertaken by the Organization for Economic Cooperation and Development and by the Atlantic Institute have shown that 40 percent of the investment in the three largest countries (West Germany, Britain, and France) was provided by Standard Oil, General Motors, and the Ford Motor Company; that another 25 percent originated from only twenty American corporations; that 460 of the 1000 largest United States corporations has bought subsidiaries or established branches in Europe in 1961, and that four years later (at a cost of $10 billion) the figure had risen to 700 out of the top 1000.

Several economists have warned that Europe could eventually lose control over its newest and most productive industries if this trend of acquisition or manufacturing licensing should persist. They have cited the experience of what United States investment has already accomplished in Canada. Though American capital and management have helped modernize and expand Canada's manufacturing capacity, the Canadians have paid a high price for their economic boost. Forty-three percent of Canada's industrial assets (including 54 percent of its capital in mining and 64 percent in oil) have passed into American hands, and many basic decisions about Canada's future are now beyond its control. A few decisions to close down or to relocate redundant plants in Europe have reinforced anxieties. In some cases, American management, by imposing abrupt changes in the pattern of economic development, showed no sensitivity to local sentiment or to national needs. Ford and General Motors both reacted to intransigence in the French auto industry by building valuable new plants in Belgium rather than in northern France. If these instances were to multiply, resentment against America's industrial ascendancy could rapidly intensify. Europe rightly fears that the distribution of economic power could be determined by decisions taken in the board rooms of New York or Chicago rather than in the planning commissions of Europe. Indeed, the pressures exerted by our State Department on behalf of Britain's entry into the Common Market were sufficiently aggressive to remind both left- and right-wing parties of the influence which might be brought to bear upon their fragile strength if the United States should ever be crossed.

The impact of America has forced a rapid change in Europe's social and cultural values. As concern with economic growth began to command attention, European industry discovered that there was an alarming shortage of trained men and that most of the educational institutions throughout Europe were impossibly rigid in their technical curricula, deficient in research training, and far too limited in capacity to produce skilled graduates in adequate numbers. The advanced nations of Western Europe have spent less than half of the sum which the United States has invested in improving the per capita productivity of its work force. While 40 percent of college-age youth in the United States enter some form of higher education, the proportion in Europe is barely 10 percent. As a result, there are four or five times as many qualified scientists and engineers in the New World as in the Old and productivity per man is just about double. Though the supply of well-trained young men is relatively greater in the United States, the needs of American industry still remain unfilled. United States or multinational corporations in Europe thus recruit from the European minority with university or technical education. Salary and research facilities exceed what European employers can afford, and the emigration of skilled men became conventionally known as "the brain drain."

Numerous complaints are made about this pirating. The annual migration of 13,000 professional or technical staff has deprived the Old World of scarce and expensively trained scientists and it has been actively encouraged to the dismay of many European and Third World nations, by the discriminatory clauses in American immigration laws. Officials in Britain have estimated that it costs $28,000 to train a competent scientist or physician and that 37 percent of all Doctors of Philosophy produced by physics faculties between 1958 and 1963 had been lured away after their education had been paid for at home. The emigration of staff from British aircraft plants or medical research centers testifies to the superior attractions that science graduates have found in the American corporations that subsidize research and seriously consult with their technical staff. Many of the émigrés encountered these attractions for the first time in British companies working under license or control to American management. In many cases their transference across the Atlantic followed soon afterward.

European frustration and resentment have increased proportionately with the scope of the American industrial invasion. Political leaders in West Germany and Britain realize that there is little that they can do to suppress this profitable but unbalanced modernization of economic capacity and they entertain few illusions that our Government will take action to arrest it. But a profound ambivalence has spread across the Continent as a consequence of this new, trans-Atlantic form of colonialism. Though the inflow of American capital and technology has generally been welcomed, even by left-wing parties, the meddling of American parent firms in national planning priorities is a problem. For example, the United States Government instructed a French subsidiary of the Fruehauf Trailer Company to cancel an order to deliver trucks to China. The decision of the French government to seize the subsidiary and cancel the instruction indicates that there may be an abrasive pattern in trans-Atlantic relations in future years.

Ill-feeling toward the United States has tended to develop into a serious problem in Western Europe as fears of a technological lag continue to multiply. Speaking to an assembly of European leaders in Strasbourg, for instance, Prime Minister Wilson capitalized on the mood of apprehension and fear prevailing among Britain's unfriendly neighbors. He warned that Europe must not become "an industrial helot to the sophisticated apparatus of American business," and he urged that a unified European Technology Community should be established to compete with America's industrial superiority. A more striking suggestion, to integrate Europe's industrial growth with the aid of a technological Marshall Plan, was made by the Italian Foreign Secretary, but it failed to impress our Government. Across the Channel, the warnings of the Gaullists were more strident. In his recent book, *The Paradoxes of Peace,* General Gallois accused the United States of planning "the technological conquest of Europe through fifth column methods of industrial subjugation." Objecting strenuously to America's nuclear monopoly in Europe, to its yearly arms sales of $2 billion on the Continent, and to its devious promotion of Britain's entry into the Common Market—to serve, no doubt, as a Trojan horse for America's industrial penetration—he charged that United States policy sought to keep Europe in a condition of permanent underdevelopment.

De Gaulle explained the last French veto of Britain's admission to the

Common Market on the grounds that the economic autonomy of a *Europe des patries* must be preserved from the technological mastery of the Anglo-Saxons. In the next year he accelerated his anti-American attack by forcing a reappraisal of world gold values and thus of the parity price of the dollar. He complained that the United States had exported the domestic inflation and budget deficit resulting from the excesses of the Vietnam war by forcing Europe temporarily to support the $4 billion imbalance of payments which the United States had incurred with Western Europe. His assault upon the dollar severely disturbed the complacent views held on Wall Street and in the United States Treasury, but it was eventually doomed to fail. The International Monetary Fund and the European central banks were too determined to preserve the stability of their international reserve currencies to succumb to French pressure. A few months later France was crippled by a general strike and ominous inflationary pressures, and the franc lost value as the dollar regained its strength and as the West Germans refused to increase the value of their own strong currency. Though the Gaullist assault petered out when the franc was devalued by 12.5 percent to shore up the French economy, the assault had taught several useful lessons. First, it indicated how numerous were the sympathizers with the French argument that the financial control and investment needs of Europe should not be ceded to American direction. Second, it demonstrated that the advanced economies of the West were vitally dependent on each other—and more so upon the United States—to maintain the delicate balance between inflationary growth and deflationary recession. Third, it was shown that the extensive power of American industry, though not invulnerable, could basically determine the pace of change and the mood of confidence in Western Europe.

The efforts of several nations to bridge the widening gap and to invigorate Europe's industries of high technology have been jointly planned in the field of space research, computer and nuclear engineering, in electronics, and in tele-communications; but the progress reported so far has been pitiably inadequate. Most of the nations of Western Europe are still unwilling, for political reasons, to unify their scientific work, to coordinate their plans for industrial expansion, and to commit themselves to a cost-effective division of labor. Their inability and unwillingness to integrate growth plans have impeded the modernization of Europe's industrial base and have thus allowed United States corporations to establish themselves firmly in many of Europe's most lucrative markets. Tragically, the Europeans have realized that they will never be able to close the gap until they integrate their efforts—but they have also recognized that the historic obstacles which prevent the unification of Europe will survive for at least another generation

The threat that Europe faces today should not be underestimated. The valuable results of America's spending in technology are only now beginning to appear in the "multiplier" effects of increased productivity, and greater benefits will become visible in the 1970's. If the rate of innovation and economic growth in the United States should reach new heights, Europe's lag will assume catastrophic dimensions. It is likely that the annual GNP of the Unites States will soon exceed a trillion dollars and that its expenditures in R & D will vastly extend the forcefulness of America's technological leadership. Though three-fifths of America's R & D expenditure is deployed for military purposes, it is now evident that more industrial research is conducted in California alone than in the whole of Western Europe. The United States invests well over $100 per capita or 4.3 percent of its annual national income in R & D; by contrast, Britain invests 2.9 percent, France 2.1, and West Germany 1.8. The product of this investment will become visible in the 1970's as the United States begins to reap the profits of technological innovation which Europe can no longer hope to match. As Servan-Schreiber put it, unless the Continent can unify its modernization programs and discard its archaic forms of economic nationalism, the combined industrial product of Western Europe in the late 1970's will only be as great as the United States national product of today.

Further American control over Europe's high technology industries will assuredly be acquired if the gap between the productive skills and capacities of the New World and the Old should continue to grow. United States plants have invested in the sophisticated apparatus of lasers, nuclear reactors, automated engineering, holography, and solid-state physics. Their subsidiaries and licensees in Europe have already achieved a higher return than many of their European competitors by devising new methods of management and marketing; and they have been notoriously more successful in raising the capital—even in Europe—needed to finance an enlarged production schedule. Were they to acquire an even larger slice of Europe's industrial market, no imaginable grouping or economic aggregation of power in Western and Eastern Europe could then close the technology gap.

With these facts in mind, an attempt can now be made to define what is meant by the technology gap. It can be defined initially as a trans-Atlantic *disparity* in the growth rates of technological innovation, educational develop-ment, and entrepreneurial skills. It can also be identified with the *discrepancies* between the per capita productivity, the technical use of capital, and the ability to generate a faster rate of return that distinguish American from European industrial performance. These discrepancies do not result simply from the inequality of GNP or per capita income. The case of Japan reveals that technical and educational progress is more important for development purposes than the mere enlargement of GNP. More to the point, an acceleration in the acquisition of technical knowledge or in the capitalizing of new technological processes is vital if output-per-man is to increase and if production costs are to be radically reduced. It is for this reason that the brain drain becomes so painful a burden to societies struggling to modernize their work force. It is for this reason, too, that

Europeans fear that the United States will maintain its patents and royalties upon the new skills and processes of production, leaving it to the Europeans to run the less profitable industries which the United States ignores.

Europe is now in a position of marked if not permanent inferiority before the vast growth of American skills and resources. Were the deflection of wealth and energy to the war in Vietnam and to the swollen American defense budget to cease, the disparity between Europe's and America's industrial potentials would be even more impressive. Perhaps because they are embittered by the towering advantages of the United States, or perhaps because they resent the uses to which it has put its wealth in controlling world affairs, European leaders have begun to complain about the chaotic educational and social pressures which have surfaced in the United States. They disdain our materialistic and inhumane values; and they condemn—with some justice—the endless financing of defense and its industrial spin-off benefits while the welfare and public sector needs of the economy remain impoverished. As the serious press in London, Paris, Zurich, or Frankfurt has shown, there is a widespread mixture of envy of us and contempt for our troubles in mass universities, for the impersonal assembly work in our industrial centers, and for the impossible conditions within our riot-torn cities. Following the social disorders that erupted in our racial ghettos and in our "prisons of poverty," as Le Monde put it, there was outspoken condemnation of the human costs which we have paid to pursue a full-scale war together with a high growth rate, simultaneously.

But more significant than the recrudescence of anti-American sentiment is the despair among those who cherish the traditional values and the settled arrangements of European society. There can be no place for the European spirit, they have argued, in a future characterized by the forced specialization of work and the technical re-ordering of society. Labor unions, university professors, and even bankers have been adamant on this point. Europe's submission to America's nuclear and military domination required twenty years of bitter adjustment, they point out; proud colonial possessions had to be abandoned, the ideological cleavages among political parties ignored, and the diplomatic leadership of Europe historically foresworn. To now submit to an American ascendance in technical know-how and industrial expansion is a humiliating prospect. It might require another two decades before their civilized protest is stilled and the long travail of industrial adjustment is finally consummated.

The inability of Europe to accelerate its rate of technological growth has been demonstrated. It has been impossible for Europeans to build the unified institutions that they most desperately require: an integrated mode of production which crosses national lines, a transnational division of labor to encourage the fullest mobility of manpower, a single-structure market which would allow Western Europe to enjoy the advantages of large-scale and specialized production. While it took the United States two or three hundred years to achieve a thorough integration of its productive and distributive mechanisms, this mode of economic rationalization has been historically closed in Europe. Today the goal of building a United States of Europe, though long urged by Jean Monnet and other advocates of a federated and functional unity, is still dismissed as a utopian vision. The six nations of the European Economic Community cannot agree whether Britain and its European Free Trade Association colleagues should be admitted into the Common Market, whether industrial operations should be allowed to merge across national boundaries, and whether trade and investment planning should be undertaken on a regional basis. Most of the governments of Western Europe still regard their science-based industries as fragile national symbols rather than as vital contributors to a European-wide economy. The sense of patriotism in France has continued a needless rivalry with British industry and a sustained suspicion of West Germany. But more than any other factor, the persistence of political divisions has maintained Europe's economic inferiority and fragmentation. The American computer or aircraft industries now service a market (at home and overseas) that is three or four times larger than all of Europe's markets combined. Yet so long as British, French, German, and Italian firms insist upon preserving their own markets and national production schedules, the ability of any one firm to compete with an American giant is effectively precluded.

The dilemma that Europe faces is not enviable. The enormous benefits that could flow from American investment and know-how have at least been recognized. These include the development of new production techniques, research and management controls, the payment of higher wages, the geographical relocation of industry, the subsidizing of technical and educational institutions, and the restructuring of the tax base of the economy. But there is also a widespread fear that the sovereignty of each nation will be further impaired as one economy becomes more interdependent with those of its neighbors. The distrustful nations of Western Europe must therefore choose whether to return to a mercantilist theory of shutting out foreign competition and capital, or whether to search for a new form of international cooperation to reduce the burden of technical lag. Ironically, there is one change which they must dare undertake immediately. Instead of continuing to invest surplus capital in highly diversified portfolios in the United States stock market, they must utilize their funds either to finance their own growth projects or to buy control over parent corporations within the United States—as British petroleum firms have done while seeking a position in the Alaska oilfields.

At present there is no clear resolution for the dilemma. Most schemes for industrial mergers and cross-national cooperation appear to threaten the survival of the science-based industries of the Rhone, the Ruhr, and the Mersey valleys. Though integration and mergers across national lines in Europe are desperately needed, they cannot begin until the European Six and Seven agree upon future goals of political unification. (Among these goals, it should be noted, is a growing desire to include integration with the non-market economies of Eastern Europe, many of which can offer opportunities that statesmen in Paris or Bonn are reluctant to ignore.) No matter what goal priorities are finally selected, the process of modernization will require cooperation rather than conflict with American firms and their numerous subsidiaries. On the American side, corporations such as Standard Oil and the Bank of America have already begun to experiment with international boards of directors and with multinational planning exercises. But even these experiments will fail if United States firms continue to dominate lucrative markets or to build American enclaves within the growth sectors of each economy. Political change in Europe at the present time has been limited largely to tariff manipulations or to the securing of agricultural price supports. If Europe does not make revolutionary changes, it can be safely predicted that United States management will seize the industrial advantages which the Europeans refuse to exploit for their own purposes.

It is now obvious that the most advanced nations of Western Europe can afford neither to duplicate the costly technology of the United States nor to shut out its beneficial consequences. A few nations—such as Sweden and the Low Countries—have chosen a parochial but successful strategy of concentrating their growth efforts in critical industries while relying upon American assistance and licensing in others. It remains to be seen whether their larger neighbors can bring themselves to follow this pattern of partial or semisovereign development. Though Japan or Holland has been able to adopt this model of economic planning, it is improbable that Britain, France, or West Germany will do so. In each of the major "middle powers" there is an element of pride that cannot be discarded. Each had formerly been an industrial pioneer in establishing its own pattern of affluence, empire, and growth. None is able or willing to adapt quickly to a rationale of limited, interdependent, and regional schemes for economic development.

No matter which model of development is pursued in Western Europe, there are three basic requirements for significant industrial progress. First, each nation bent upon industrial modernization and expansion must revolutionize the social recruitment, education, and deployment of its scientific and managerial staff; to program the phasing-in of automated production with a class-bound elite or a poorly trained work force is obviously absurd. Second, a large group of European nations must commit themselves to a set of shared goals larger than the concessions so far legislated by the Common Market; these must include the elimination of small and unproductive industries, the merging of research and of marketing operations, and the removal of wasteful forms of national competition for scarce resources. Third, the Continent must devise unified political institutions to accelerate the pace of technological innovation and economic integration; thus if new institutions (like Euratom) should function smoothly they will overshadow and then replace the national rivalries which still divide the skilled populations of Europe, but if they should remain weak and distrusted the phenomena of technical lag will surely persist.

Unfortunately, the political leadership of a post-de Gaulle Europe does not appear committed to a revolutionary program of social and international change. It falsely believes that time can still be bought—with instrumental modifications such as the Kennedy Round of tariff reductions or the consolidation of international reserve currencies—in order to delay the onset of rudimentary change. Hence, the scientific research centers, the national planning commissions, and the labor management institutions of Western Europe more closely resemble the Europe of 1939 than the Europe which should emerge in the last quarter of this century.

Though the prospect of an unchallengeable industrial superiority might appeal to the more complacent or profit-minded apologists for American policy overseas there is good reason for future concern. A frustrated Germany, resenting its exclusion from the privileges of nuclear power and covetous of its weaker neighbor to the East, will not indefinitely stifle its national ambitions with American licensing arrangements or French price-fixing plans. Nor will a divided Continent ever discard its dangerous reliance upon the stability born of armed confrontation if it must resign itself to a satellite industrial status. More important, Europe will be unable to fulfill an adequate role in the international system if it turns inward upon itself in frustration and despair in the 1970's. Without question, the industrial gap between North American and Europe must be narrowed if political stability and the promise of growth are to be assured. Once this process has begun, the truly desperate gap—in the sustenance and life opportunities between the northern and southern hemisphere—can at last be filled and the awesome specters of mass starvation and upheaval laid to rest. If Europe is neither powerful nor confident enough to assist in this most compelling task, the United States will become the target for most of the free-floating hostility and aggression of a thwarted world.

Bibliography

For the student who is interested in examining the topic at greater length, the following bibliography has been prepared. It is selective rather than exhaustive, and the writings have been chosen on the basis of their breadth and vividness of coverage, their readability, and their provocative and controversial nature.

A work which illustrates quite sharply the relevance—verily, the clear necessity—of analyzing the past in order to understand the present, is Leo Marx's elegant study, *The Machine in the Garden*. Professor Marx points out that historically, Americans have been torn between their idealization of the wilderness and the frontier (the garden) and their fascination with the technology (the machine) which destroys that wilderness.

The concept of the "Technetronic Era" or "Post-Industrial Revolution" era is developed and analyzed in the following works:

Jean-Jacques Servan-Schreiber, *The American Challenge* (New York, 1968).

Zbigniew Brzezinski, *Between Two Ages: America's Role in the Technetronic Era* (New York, 1970).

Daniel Bell, "The Post-Industrial Society," in Eli Ginzberg, ed., *Technology and Social Change* (New York, 1964).

Richard P. Schuster, ed., *The Next Ninety Years* (Pasadena, 1967).

John McHale,*The Future of the Future* (New York, 1969).

R. Buckminster Fuller, "Planetary Planning", *The American Scholar,* Winter 1970-71, Spring 1971.

245

Special editions of learned journals and other periodicals are often helpful in introducing one to a broad spectrum of authors interested in the Technological Revolution. Good examples are:

The American Scholar
   "The Electronic Revolution", Spring, 1966.
   "Youth 1967: the Challenge of Change", Autumn, 1967.
Daedalus
   "Science and Culture", Winter, 1965.
   "Toward the Year 2000: Work in Progress", Summer, 1967.
   "The Conscience of the City", Fall, 1968.
Current
   "Technology and Social Change: A New Critique of the Scientific Establishment," June, 1970.
Encounter
   "The American Crisis," January, 1970.
   See especially the ironical essay by Harold Orlans.

Among the multitude of works which express concern over the dehumanizing effects of technology, the following are useful and interesting:

Alvin Toffler, *Future Shock* (New York, 1970).
John McDermott, "Technology: The Opiate of the Intellectuals," *The New York Review of Books,* July 31, 1969.
Paul Goodman, "Can Technology be Humane?" *The New York Review of Books,* Nov. 20, 1969.

Even in such a limited listing, it would be impossible to ignore the monumental volumes on man and technology created by Lewis Mumford:

Technics and Civilization (New York, 1934).
The City in History (New York, 1961).
The Myth of the Machine: Technics and Human Development (New York, 1967).
The Myth of the Machine: The Pentagon of Power (New York, 1970).

An interesting analysis of the work of Lewis Mumford and Buckminster Fuller is contained in:

Allan Temko, "Which Guide to the Promised Land: Fuller or Mumford?" *Horizon,* Summer, 1968.

A good survey of the problems and perils involved in attempting to save the environment is found in Raymond F. Dasmann, *A Different Kind of Country* (New York, 1968). A challenging concept elaborated by Mr. Dasmann is "planning against progress." Another problem looming on the horizon is analyzed by scientific writer Ralph Lapp in "Power-Hungry America: Where Will We Get the Energy," *The New Republic,* July 11, 1970. The issue of the population explosion is handled vividly by Paul and Ann Ehrlich in *Population, Resources, Environment* (San Francisco, 1970). A good introduction to the complex issue of science and government is Edward T. Chase's essay, "Politics and Technology," *The Yale Review,* Spring, 1963.

While the literature on the Youth Movement is vast, several essays relate this subject directly to the Technological Revolution. Charles Reich, *The Greening of America: How the Youth Revolution Is Trying to Make America Livable* (New York, 1970) and Theodore Roszak, *The Making of a Counter Culture: Reflections on the Technocratic Society and Its Youthful Opposition* (New York, 1969), are popular, flamboyant, and passionately sympathetic. A pair of provocative articles which present diametrically opposed interpretations is:

John H. Schaar and Sheldon S. Wolin, "Where We Are Now," *The New York Review of Books,* May 7, 1970, and Bruno Bettleheim, "Obsolete Youth . . . . ," *Encounter,* September, 1969.

In surveying the impact of technology on the global scene, an interesting introduction to the interrelated subjects of industrial development and East-West "convergence" is provided by Raymond Aron in *The Industrial Society* (New York, 1968). The following are readable and controversial works about the industrial and technological super-powers—Japan, U.S., and U.S.S.R.:

Herman Kahn, *The Emerging Japanese Superstate, Challenge and Response* (Englewood Cliffs, N.J., 1970).
Andrei D. Sakharov, *Progress, Coexistence, and Intellectual Freedom* (New York, 1968).
Andrei Amalrik, *Will the Soviet Union Survive Until 1984?* (New York, 1970).
Jack N. Behrman, *Some Patterns in the Rise of the Multinational Enterprise* (Chapel Hill, 1969).
Charles P. Kindleberger, *American Business Abroad: Six Lectures on Direct Investment* (New Haven, 1969).
The Technology Gap: U.S. and Europe (New York, 1970); prepared by the Atlantic Institute.

Index

249